CRC Handbook of Lichenology

Volume III

Editor

Margalith Galun, Ph.D.
Professor
Department of Botany
The George S. Wise Faculty of Life Sciences
Tel Aviv University
Tel Aviv, Israel

CRC Press is an imprint of the
Taylor & Francis Group, an **informa** business

CRC Press
Taylor & Francis Group
6000 Broken Sound Parkway NW, Suite 300
Boca Raton, FL 33487-2742

Reissued 2019 by CRC Press

A Library of Congress record exists under LC control number:

Publisher's Note
The publisher has gone to great lengths to ensure the quality of this reprint but points out that some imperfections in the original copies may be apparent.

Disclaimer
The publisher has made every effort to trace copyright holders and welcomes correspondence from those they have been unable to contact.

ISBN 13: 978-0-367-26179-5 (hbk)
ISBN 13: 978-0-367-26180-1 (pbk)
ISBN 13: 978-0-429-29186-9 (ebk)

**Visit the Taylor & Francis Web site at http://www.taylorandfrancis.com and the
CRC Press Web site at http://www.crcpress.com**

PREFACE

Lichens are among the most widely distributed eukaryotic organisms in the world. There are, to date, about 13,500 lichen species known, which accounts for approximately 20% of all the fungi described. All are the result of a symbiotic association between two unrelated organisms — a fungus and an alga (or cyanobacterium) — which, when fully integrated, form a new biological entity with very little resemblance to either one of its components.

Despite their wide ecological amplitude and abundance, also in extreme environments where often other plants cannot exist, lichens have received relatively little attention by plant physiologists. It was not until 1984 that the first international conference on lichen physiology and cell biology was held (organized by D. H. Brown, in Bristol). The second conference, devoted in part to lichen physiology, took place in 1986 (organized by E. Peveling, in Münster).

I have attempted to invite as many as possible lichenologists who have made significant contributions to lichen research, to contribute chapters to this book. Each author has been encouraged to approach his subject in his own way, in order to communicate the author's enthusiam to the reader. Naturally, in a book where each chapter deals with its own topic, there is bound to be some overlap of subject matter. Thus, people who are interested in a specific subject and read only one of the chapters, need to have some background pertinent to that subject. The reader will also find some contradictory statements in the various chapters, mainly on assumptive aspects, which I did not find necessary to unify, in order to stimulate further investigations.

Topics which have not been comprehensively surveyed in other treatises or are scattered in different journals, such as Nitrogen Metabolism (Chapter VI.B), Enzymology (Chapter VI.C), and Ecophysiological Relationships in Different Climatic Regions (Chapter VII.B.2), are here given more substantial scope. The Algal Partner (Chapter II.B) is much more extended that The Fungal Partner (Chapter II.A), because the systematic status of the photobionts has not been treated before in an integrated manner.

The terms "phycobiont" for the algal symbiont and "mycobiont" for the fungal symbiont of lichens have been introduced by G. S. Scott in 1957 [Lichen Terminology, *Nature (London),* 179, 486, 1957]. After the "blue-green algae" were classified as "cyanobacteria" it was proposed (Ahmadjian, V., Holobionts Have More Parts, *International Lichenological Newsletter,* 15, 19, 1982) that "photobiont" should be used for the photosynthetic partner(s) of the lichens and more specifically "cyanobiont" for the cyanobacterial symbionts and "phycobiont" for the eukaryotic algal symbionts. This terminology has been adopted here, except for Chapter II.B (Tschermak-Woess) where, upon the author's request, the previous terms "phycobiont" and blue-green algae (Cyanophyceae) appear.

The names of the lichens cited in the various chapters are always those of the original articles. This may, in many cases, not coincide with the modernized system by Haffelner (Chapter X).

It is my hope that this book will bring lichenology to the notice of biology researchers, teachers, and students, as an important phenomenon in the mainstream of biology.

I greatly appreciate the efforts of the contributors and am especially indebted to Dr. Paul Bubrick, who encouraged me to launch this endeavor and who was very helpful in the editing of several chapters. I also wish to thank Dr. Leslie Jacobson, who read and commented on the chapters I contributed to this book, and Ms. Ruth Direktor and Ms. Henriette van Praag for their skillful secretarial assistance.

Finally, I am deeply grateful for the help, forbearance, and encouragement of my family.

Margalith Galun

THE EDITOR

Margalith Galun, Ph.D. is Professor of Botany at The George S. Wise Faculty of Life Sciences of Tel Aviv University. She received her M.Sc. and Ph.D. degrees from The Hebrew University of Jerusalem. Dr. Galun is a member of several professional and scientific organizations and in some she holds executive positions, such as Vice-President of the International Association for Lichenology, while in others she serves on the Executive Committee, such as that of the International Mycological Association and of the Israel National Collections of Natural History. She is the editor-in-chief of the journal *Symbiosis* and is a member of the editorial board of the journals *Israel Journal of Botany, Endocytobiosis and Cell Research,* and *Lichen Physiology and Biochemistry*. Dr. Galun has been the recipient of many research grants and is the author or co-author of two books and of about 90 research articles, including reviews, symposia, and chapters in books. Her current major research interests relate to the interaction between symbionts of plant symbiotic systems.

CONTRIBUTORS

Volume III

Madalena Baron, M.Sc.
Professor
Departamento de Bioquimica
Universidade Federal do Parana
Curitiba, Parana, Brazil

Paul Bubrick, Ph.D.
Research Associate
Department of Plant Pathology
University of California
Riverside, California

Bazyli Czeczuga, Sc.D.
Professor
Department of General Biology
Medical Academy
Bialystok, Poland

Margalith Galun, Ph.D.
Professor
Department of Botany
The George S. Wise Faculty of Life Sciences
Tel-Aviv University
Tel-Aviv, Israel

Philip A. J. Gorin, Ph.D.
Professor
Departamento de Bioquimica
Universidade Federal do Parana
Curtibiba, Parana, Brazil

Josef Hafellner, Ph.D.
Universitäts Dozent
Department of Botany
Karl Franzens University
Graz, Austria

Marcello Iacomini, Ph.D.
Professor
Departamento de Bioquimica
Universidade Federal do Parana
Curitiba, Parana, Brazil

Adiva Shomer-Ilan, Ph.D.
Senior Lecturer
Department of Botany
The George S. Wise Faculty of Life Sciences
Tel-Aviv University
Tel-Aviv, Israel

John L. Innes, Ph.D.
Research Scientist
Forestry Commission
Alice Holt Lodge
Wrecclesham
Farnham, Surrey, England

David Jones, Ph.D.
Principal Scientific Officer
Department of Microbiology
Macaulay Land Use Research Institute
Craigebuckler, Aberdeen, Scotland

David H. S. Richardson, D.Phil.
Professor
School of Botany
University of Dublin
Trinity College
Dublin, Ireland

Reuven Ronen, Ph.D.
Research Associate
Department of Botany
The George S. Wise Faculty of Life Sciences
Tel-Aviv University
Tel-Aviv, Israel

TABLE OF CONTENTS

Volume I

TABLE OF CONTENTS

Volume II

TABLE OF CONTENTS

Volume III

Section IX: Chemical Constituents of Lichens
IX.A. Secondary Metabolic Products
IX.B. Storage Products of Lichens
IX.C. Pigments

Chapter IX.A

SECONDARY METABOLIC PRODUCTS

Margalith Galun and Adiva Shomer-Ilan

I. INTRODUCTION

The secondary metabolites of lichens, often referred to as "lichen substances" or "lichen acids", are one of the more intensively investigated aspects in lichenology. They are routinely used in lichen systematics, more so than chemical substances, of any other group of organisms.

The "chemistry" of about one third of all lichen species has now been studied and about 350 secondary compounds are known from lichens. Many of these substances are entirely specific to lichens and are synthesized as the result of the symbiotic conditions. The chemical structure of approximately 220 of them has been established, and continually more and more are chemically characterized.

They are extracellular products of relatively low molecular weight, crystallize on the hyphal cell walls, are usually insoluble in water, and can be extracted only with organic solvents. They amount to between 0.1 and 10% of the dry weight of the thallus, sometimes up to 30%. One species may contain one to several different substances. Some are colored and others are colorless.

Lichen substances have a history over many centuries attributable to their therapeutical and other uses (see Chapter XII.B).

Secondary metabolites are generally considered " . . . compounds that have no recognized role in the maintenance of fundamental life processes in the organism that synthesize them".[1] However, many investigations tried to attribute biological functions to the lichen substances. The pigmented cortical compounds have been interpreted as having a role in protecting the photobiont from too much light. Lichens have a long life span (see Volume II, Chapter VII.A), and the secondary products are considered to have defensive functions against parasitizing fungi or bacteria and browsing animals. Lichen substance crystals are water repellent and have, therefore, been considered functional in gas exchange of hydrated thalli. It has to be noted, however, that not all lichen species contain "lichen substances"; mainly those with cyanobionts are lacking in these substances.

II. CLASSIFICATION

The secondary lichen metabolites are very heterogeneous and belong to a variety of chemical groups (see Figure 1):[2-7]

- Acetate-polymalonate group
  - Higher fatty acids and lactons
  - Aromatic polyketides
    - Depsides
    - Depsidones
    - Dibenzoquinones
    - Dibenzofurans
    - Chromones
    - Xanthones
    - Naphtoquinones
    - Anthrones
    - Anthraquinones
- Mevalonate group (see Chapter IX.C)

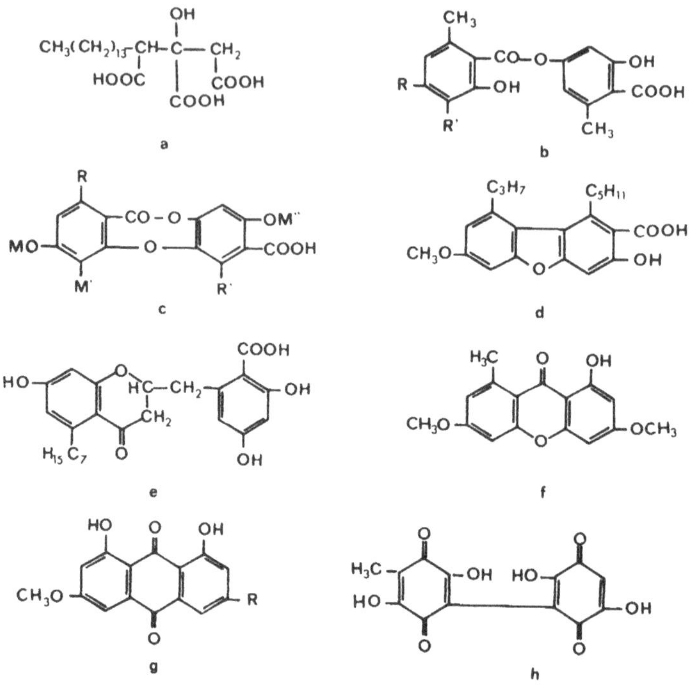

FIGURE 1. Structural formulas of some lichen substances: (a) caperatic acid (an aliphatic acid); (b) diploschistesic acid (a depside), R,R′ = OH; (c) alectoronic acid (a depsidone), R,R′ = $CH_2COC_5H_{11}$ M, M′, M″ = H; (d) didymic acid (a dibenzofuran); (e) siphulin (a chromone); (f) lichexanthone (a xanthone); (g) parietin (an anthraquinone), R = CH_3; (h) chiodectonic acid (a napthtoquinone).

Shikimate group
- Diketopiperazines
- Terpenylquinones
- Pulvinic acid derivatives

The aromatic polyketides are the most common lichen products. Depsides, depsidones, dibenzoquinones, and dibenzofurans are found exclusively in lichens. The other aromatic polyketides are not unique to lichens and have also been detected in nonlichenized fungi and higher plants.

III. DETECTION AND IDENTIFICATION

A. Color Tests

Color tests were invented in 1866 by Nylander,[8] even before the dual nature of lichens was recognized (see Volume I, Chapter I), and have been used ever since in taxonomic determinations. He introduced potassium hydroxide (K) and calcium hypochlorite (C). Paraphenylenediamine (PD) as reagent was added at a later stage.[9] (For the accurate preparation of these reagents, the reader is referred to Hale.[10])

The reagents have to be applied (separately) on thallus fragments (on cortex and medulla separately) and the change of color observed immediately. Apothecia often react differently than the thallus:

K: reacts with quinoids, depsides, and β-orcinol depsidones
C: reacts with aromatic compounds that have two free meta-hydroxyl groups
K followed by C: reacts with orcinol depsidones

Color reactions are mentioned in most key books as additional characters for lichen identification.

B. Microcrystal Tests

Between 1936 and 1940, Asahina[11] devised simple microcrystallization tests for lichen substance identification which can be performed on microscope slides. A few lichen fragments are crumbled on a microscope slide and extracted with a few drops of acetone (or other organic solvents if required). After evaporation of the solvent, the fragments are removed gently (with a soft brush) leaving a residue of crystals on the slide. A drop of a recrystallizing reagent is added to the residue and a coverslip is placed directly over it. The slide is heated gently over an alcohol lamp or an electric plate. Upon cooling, the crystals that appear and have distinctive shapes are examined microscopically and can be identified by comparison with photographs[12-15] of the specific substances (see Figure 2). The most common recrystallizing reagents are GE, glycerol:acetic acid (1:3); GAW, glycerol:ethanol:water (1:1:1); GAoT, glycerol:ethanol:o-toluidine (2:2:1); and GAQ, glycerol:ethanol:quinoline (2:2:1).

C. Thin-Layer Chromatography (TLC)

TLC is a more accurate and very common technique for detecting and identifying lichen substances. Optimal conditions have been developed in the laboratory of Culberson[16-20] and are recommended as standard procedures. The lichen samples are extracted in toluene and acetone and the extracts then loaded on Merck Silica gel 60 F_{254} plates and chromatographed in three standard solvent systems: (1) toluene-dioxane-acetic acid (180:45:5, 230 mℓ); (2) hexane-diethyl ether-formic acid (120:90:20, 230 mℓ); and (3) toluene-acetic acid (200:30, 230 mℓ). Methyl *tert.*-butyl ether (MTBE) recently has been introduced to substitute diethyl ether in the second solvent if large numbers of samples are examined routinely, because MTBE is less hazardous.[21] Several additional solvents have been suggested for specific cases.[19,20] The spots on the chromatograms are either visible in normal light or have to be viewed in short and long wavelength ultraviolet (UV) light, sprayed with 10% H_2SO_4, and heated to 110°C for 15 to 30 min to make the spots of aromatic and alicydic compounds visible. Spots of fatty acids can be located by wetting the plates in water and observing opaque spots as the plates dry on a slide-warming plate. The two common lichen substances atranonin and norstictic acid, with known R_f values, are used as control substances for relating R_f values of the unknowns examined. Hydrolysis products and O-methylated extracts are chromatographed in parallel with the crude extracts for better detection and improved distinction between substances with very close R_f values. The developing tanks have to be insulated from external temperature fluctuations.

Chromatographic data and R_f classes of 149 lichen products, 37 hydrolysis products of depsides, 24 O-methylated derivatives, and 10 miscellaneous derivatives of nondepside lichen compounds determined by TLC in the three standard solvent systems are given by Culberson.[17]

Two-dimensional TLC was devised by Maass[22-25] to determine components of complex mixtures of substances and was then modified[26] to enable correlation with data of one-dimensional TLC.

High-performance liquid chromatography (HPLC) is an analytical tool which yields quantitative measurements of mixtures of components in crude extracts, also of small samples,[27,28] which in some cases are not separable by TLC.

Spectral properties and structural elucidations of purified lichen products are possible by ^{1}H- or ^{13}C-nuclear magnetic resonance (NMR) and mass spectrometry.[29-33] The latter is particularly useful for the colored products such as xanthones and anthraquinones.

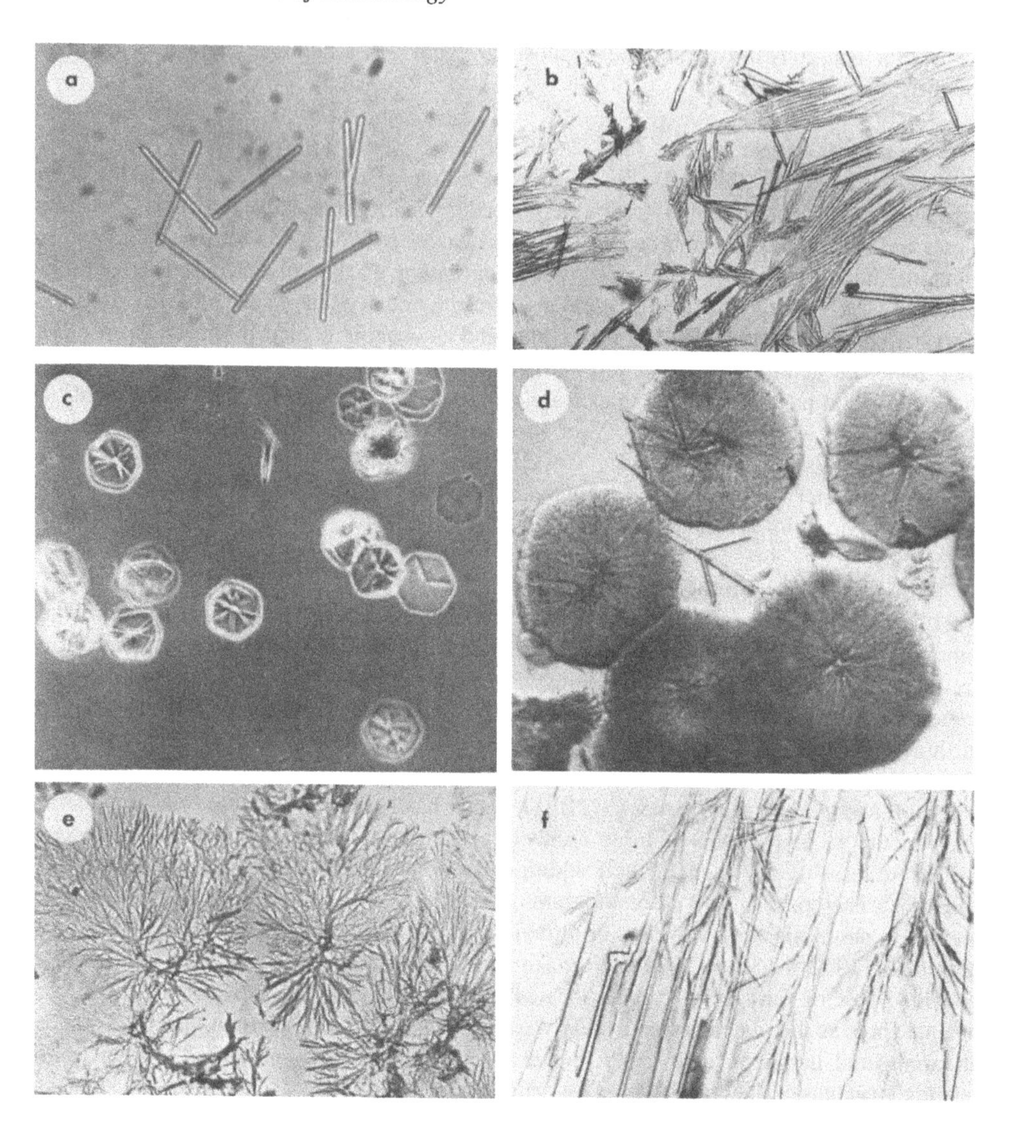

FIGURE 2. Examples of crystals: (a) atranorin, in GE, from *Physcia stellaris*; (b) usnic acid, in GAW, from *Cladonia convoluta*; (c) stictic acid, in GAoT, from *Buellia zoharyi*; (d) lecanoric acid (and atranorin), in GE, from Diploschistes actinostomus; (e) lecanoric acid, in GAW, from *D. ocellatus;* and (f) barbatic acid and psoromic acid, in GE, from *Rhizocarpon tinei*.

IV. BIOSYNTHESIS

There have been relatively few studies on the biosynthesis of lichen substances. The pathway leading to their formation suggested by Mosbach[34] is presented in Figure 3. For the available information on enzyme systems involved in the production of lichen substances, see Volume I, Chapter VI.C.

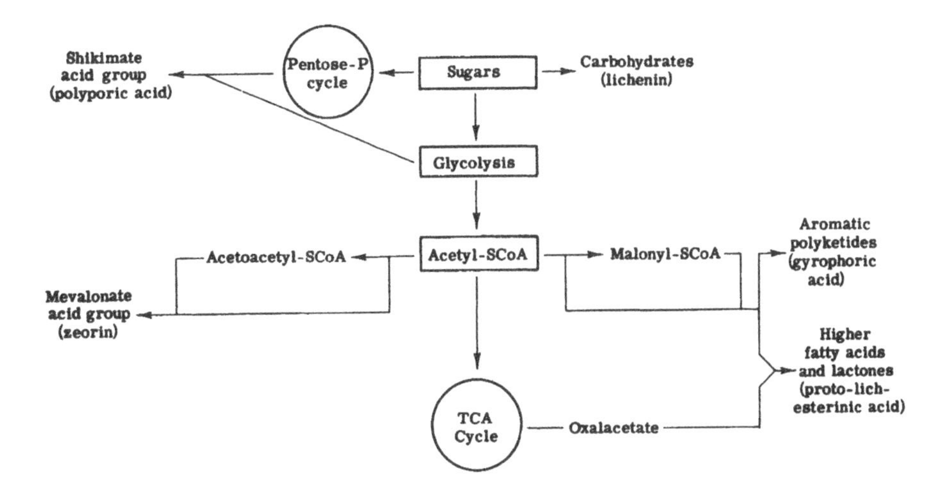

FIGURE 3. Carbon metabolism leading to the formation of the major groups of lichen substances with a typical representative given in parenthesis. (From Mosbach, K., *The Lichens*, Ahmadjian, V. and Hale, M. E., Eds., Academic Press, New York, 1973, 523. With permission.)

REFERENCES

1. **Bell, E. A.,** The physiological roles of secondary (natural) products, in *The Biochemistry of Plants*, Vol. 7, Stumpf, P. K. and Conn, E. E., Eds., Academic Press, New York, 1981, 1.
2. **Culberson, C. F.,** *Chemical and Botanical Guide to Lichen Products*, University of North Carolina Press, Chapel Hill, 1969.
3. **Culberson, C. F.,** Supplement to "Chemical and Botanical Guide to Lichen Products", *Bryologist*, 73, 177, 1970.
4. **Culberson, C. F., Culberson, W. L., and Johnson, A.,** Second Supplement to "Chemical and Botanical Guide to Lichen Products", American Bryological and Lichenological Society, St. Louis, Mo., 1977.
5. **Huneck, S.,** Nature of lichen substances, in *The Lichens*, Ahmadjian, V. and Hale, M. E., Eds., Academic Press, New York, 1973, 495—522.
6. **Santesson, J.,** Chemie der Flechten, in *Lichenes, Eine Einführung in die Flechten-Kunde*, Henssen, A. and Jahns, H. M., Eds., Georg Thieme Verlag, Stuttgart, 1979, 467.
7. **Elix, J. A., Whitton, A. A., and Sargent, M. V.,** Recent progress in the chemistry of lichen substances, *Prog. Chem. Org. Nat. Prod.*, 45, 103, 1984.
8. **Nylander, W.,** Circa novum in studio lichenum criterium chemicum, *Flora*, 49, 198, 1866.
9. **Asahina, Y.,** Uber die Reaktion von Flechten-Thallus, *Acta Phytochim.*, 8, 47, 1934.
10. **Hale, M. E.,** *The Biology of Lichens*, 3rd ed., Edward Arnold, London, 1983, 190.
11. **Asahina, Y.,** Mikrochemischer Nachweis der Flechtenstoffe, *J. Jpn. Bot.*, 12, 516, 859, 1936; 13, 529, 855, 1937; 14, 39, 244, 467, 650, 767, 1938; 15, 465, 1939; 16, 185, 1940.
12. **Hale, M. E.,** *The Biology of Lichens*, 2nd ed., Edward Arnold, London, 1974, 181.
13. **Taylor, C. J.,** *The Lichens of Ohio. Part I. Foliose Lichens*, Ohio State University, Columbus, 1967.
14. **Thomson, J. W.,** *The Lichen Genus Cladonia in North America*, University of Toronto Press, Toronto, Canada, 1967, 172.
15. **Galun, M.,** *The Lichens of Israel*, The Israel Academy of Sciences and Humanities, Jerusalem, 1970.
16. **Culberson, C. F. and Kristinsson, H.,** Standardized method for the identification of lichen products, *J. Chromatogr.*, 46, 85, 1970.
17. **Culberson, C. F.,** Improved conditions and new data for the identification of lichen products by a standardized thin-layer chromatographic method, *J. Chromatogr.*, 72, 113, 1972.
18. **Culberson, C. F.,** Conditions for the use of Merck silica gel $60F_{254}$ plates in the standardized thin-layer chromatographic technique for lichen products, *J. Chromatogr.*, 97, 107, 1974.
19. **Culberson, C. F. and Ammann, K.,** Standardmethode aur Dünnschicht-Chromatographie von Flechtensubstanzen, *Herzogia*, 5, 1, 1979.

20. **Culberson, C. F., Culberson, W. L., and Johnson, A.,** A standardized TLC analysis of β-orcinol depsidones, *Bryologist,* 84, 16, 1981.
21. **Culberson, C. F. and Johnson, A.,** Substitution of methyl *tert.*-butyl ether for diethyl ether in the standardized thin-layer chromatographic method for lichen products, *J. Chromatogr.,* 238, 483, 1982.
22. **Maass, W. S. G.,** Lichen substances. V. Methylated derivatives of orsellinic acid, lecanoric acid and gyrophoric acid from *Pseudocyphellaria crocata, Can. J. Bot.,* 53, 1031, 1975.
23. **Maass, W. S. G.,** Lichen substances. VI. The phenolic constitution of *Peltigera aphthosa, Phytochemistry,* 14, 2487, 1975.
24. **Maass, W. S. G.,** Lichen substances. VII. Identification of orsellinate derivatives from *Lobaria linita, Bryologist,* 78, 178, 1975.
25. **Maass, W. S. G.,** Lichen substances. VIII. Phenolic constituents of *Pseudocyphellaria quercifolia, Bryologist,* 78, 183, 1975.
26. **Culberson, C. F. and Johnson, A.,** A standardized two-dimensional thin-layer chromatographic method for lichen products, *J. Chromatogr.,* 128, 253, 1976.
27. **Culberson, C. F.,** High speed liquid chromatography of lichen extracts, *Bryologist,* 75, 54, 1972.
28. **Culberson, C. F., Hale, M. E., and Tonsberg, T.,** New depsides from the lichens *Dimelaena oreina and Fuscidea viridis, Mycologia,* 76, 148, 1984.
29. **Santesson, J.,** Identification and isolation of lichen substances, in *The Lichens,* Ahmadjian, V. and Hale, M. E., Eds., Academic Press, New York, 1973, 633—652.
30. **Maass, W. S. G., McInnes, A. G., Smith, D. G., and Taylor, A.,** Lichen substances. X. Physciosporin, a new chlorinated depsidone, *Can. J. Chem.,* 55, 2839, 1977.
31. **Sundholm, E. G.,** ^{13}C-NMR spectra of lichen xanthones. Temperature dependent collapse of long-range couplings to hydrogen-bounded hydroxyl protons, *Acta Chem. Scand.,* 32B, 17, 1978.
32. **Sundholm, E. G. and Huneck, S.,** ^{13}C-NMR spectra of lichen depsides, depsidones and depsones. I. Compounds of the orcinol series, *Chem. Scr.,* 16, 197, 1980.
33. **Sundholm, E. G. and Huneck, S.,** ^{13}C-NMR spectra of lichen depsides, depsidones and depsones. II. Compounds of the β-orcinol series, *Chem. Scr.,* 18, 233, 1981.
34. **Mosbach, K.,** Biosynthesis of lichen substances, in *The Lichens,* Ahmadjian, V. and Hale, M. E., Eds., Academic Press, New York, 1973, 523.

Chapter IX.B

STORAGE PRODUCTS OF LICHENS

Philip A. J. Gorin, Madalena Baron, and Marcello Iacomini

I. INTRODUCTION

The most important storage products of lichen photo- and mycobionts are free amino acids, proteins, polyols, glycosylated polyols, and polysaccharides. The best-known polysaccharides are lichenan, isolichenan, and galactomannan, each of which has a range of different, but related, chemical structures depending on the parent lichen (see also Volume I, Chapters VI.A and VI.B).

The glucans lichenan and isolichenan (and related ones of the same family) and the galactomannans, formerly called hemicelluloses,[1,2] are present in lichens in relatively high amounts and presumably arise mainly from the predominant mycosymbiont. Takahashi et al.[3] investigated the polysaccharides of six lichen species and compared them with those of phyco- and mycobionts grown in culture in terms of yield, specific rotation, solubility in water, infrared (IR) spectrum, and sugar composition. Polysaccharides of the mycobionts were shown to be virtually identical with those of the parent lichens, whereas those of the phycobionts were somewhat different. For example, the polysaccharides of *Cladonia mitis* consist mainly of mannose, galactose, and glucose, whereas those of the mycobiont contain mannose and galactose. The phycobiont, on the other hand, has two components: one is water soluble and has mainly mannose and rhamnose and the other is a water-insoluble glucan. In another case, two glucans were isolated from *Ramalina crassa*, and were soluble and insoluble in water, respectively. Similar glucans were found in the mycobiont, but the phycobiont contained a galactan whose specific rotation — 85°, suggests a β-D-furanosyl structure.

Few detailed studies have been carried out on polysaccharides of phycobionts compared to those on intact lichens. However, electron microscopy showed that the *Trebouxia* of *Physcia aipolia* contained large starch plates,[4] whereas starch was found in small discs in *Trebouxia* of *Xanthoria parietina*.[5] Starch production was highest in spring and autumn.[6] Polysaccharides of the free-living cyanobacteria *Nostoc* and *Gloeocapsa* are mucilages having complex chemical structures. That of *Nostoc* was obtained by hot water extraction and contained galacturonic acid (30%), rhamnose (10%), xylose (25%), and the remainder consisted of galactose and glucose.[7] Glycogen has been isolated from *N. muscorum* and its sedimentation coefficient determined to be 265 S.[8] The exocellular polysaccharide of *Gloeocapsa* had a specific rotation of +56° and consisted of glucose (68%), mannose (19%), and galactose (13%). A methylation-gas liquid chromatography (GLC)-mass spectroscopy (MS) study indicated nonreducing end groups (26%), 3-*O*- (17%) and 4-*O*-substituted (24%) units of glucopyranose, 2-*O*- (8%) and 3,6-di-*O*-substituted residues (10%) of mannopyranose, and 3,4-di-*O*-substituted units of galactopyranose (15%).[9]

Ozenda and Clauzade[2] expressed certainty that starch is not present in the mycobiont and that its presence has not been definitely confirmed in the algal partner. However, amylose has been encountered frequently in various species of fungi[10] as has the reserve polysaccharide glycogen. Cellulose and chitinin lichens are discussed in Volume I, Chapter VI.C and Volume II, Chapter VIII.B.

II. POLYSACCHARIDES OF LICHENS

A. Lichenan

α- and β-D-glucans with a variety of chemical structures have been isolated from lichens, and differ in the ratios of (1 → 3) and (1 → 4) linkages contained in each one. For the purposes of this review, each glucan is described structurally, but only the glucans of *Cetraria islandica* are discussed in a historical manner because of the structural questions raised and the analysis techniques used.

Berzelius[11] isolated a polysaccharide called lichenin[12] (presently called lichenan) from *C. islandica* (Iceland moss) in 1815. It was formed as a precipitate on cooling of a hot water extract. However, detailed chemical studies only commenced in 1947[13] and in these Meyer and Gürtler showed lichenan (9.5% yield) to have a specific rotation of +8°, consistent with a β configuration, and (1 → 3) and (1 → 4) linkages in a 3:7 ratio (predominant Structure 1) according to quantitative periodate oxidation data. This structure was confirmed in 1957 when Chanda et al.[14] purified lichenan by repeated solubilization in hot water followed by precipitation on cooling and precipitation of its insoluble copper complex formed with Fehling solution. In the same year, Peat et al.[15] isolated cellotriose after partial hydrolysis, showing that this structure was united by (1 → 3) links, but in 1962, Perlin and Suzuki[16] showed by enzymatic studies that occasionally three consecutive (1 → 4) linkages were present, as was indicated on characterization of the fragments formed on Smith degradation.[17]

–β–D–Glc*p*–(1→3)–β–D–Glc*p*–(1→4)–β–D–Glc*p*–(1→4)–

STRUCTURE 1.

B. Isolichenan

Meyer and Gürtler,[18] in their preparation of lichenan, found that the mother liquor contained a water-soluble glucan that could be purified by repeated freezing and thawing (0.55% yield), a process that completely removed lichenan. The product was named isolichenin (now called isolichenan) and the strongly positive specific rotation indicated α linkages. It gave a blue color with iodine which was considered to be a property of the glucan, however, in a separate investigation on another lichen, it was suggested that amylose was present in quantities insufficient for its detection by other means.[19] Isolichenan has also been purified via fractional precipitation of its water-soluble copper complex with acetone, and periodate oxidation showed that (1 → 3) and (1 → 4) linkages were present in a 60:40 ratio.[14] In contrast to the ratio obtained with lichenan, many different ratio values were determined for isolichenan, possibly partly due to difficulties in preparation of pure samples. Also using the periodate oxidation technique, the ratio was determined as 55:45[20] and 57:43.[21] In the latter study, a Smith degradation was carried out incorporating mild hydrolytic conditions and structural sequences were indicated by the formation of *O*-α-D-glucopyranosyl-(1 → 3)-*O*-α-D-glucopyranosyl-(1 → 2)-D-erythritol (43%), 2-*O*-α-D-glucopyranosyl-D-erythritol (38%) with smaller quantities of glycerol (3%), erythritol (6%), nigerose (3%), and nigerotriosyl-erythritol (5%). These values show a linkage ratio of 56.5:43.5 and, assuming the absence of amylose, indicate principal Structure 2 with some consecutive (1 → 4) links, 2 consecutive (1 → 3) links, and alternate (1 → 3) and (1 → 4) linkages, a somewhat irregular structure. ^{13}C-nuclear magnetic resonance (NMR) spectroscopy, carried out on isolichenan and prepared by successive freezing and thawing, gave another value for the linkage ratio.[19] As can be seen from Figure 1, signals at δ 82.1 and 81.9 ppm attributable to *O*-substituted C-3 and that at δ 79.3 ppm arising from *O*-substituted C-4 have an area ratio of 66:34. Clearly, standardization of methodology is necessary in determination of linkage ratios for

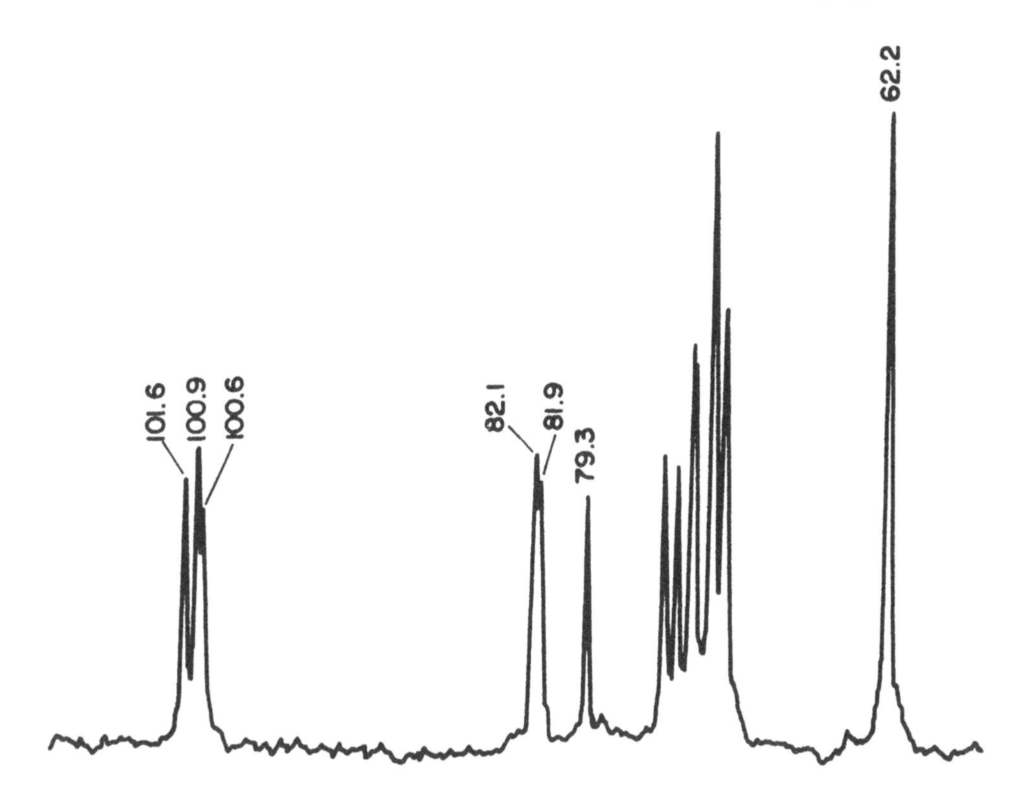

FIGURE 1. ^{13}C-NMR spectrum of isolichenan in deuterium oxide at 70° C. Numerical values are δ in parts per million.

structural comparisons within the family of glucans. Also, if impurities are present, quantitative periodate oxidation can give incorrect results. One approach is Smith degradation involving strong hydrolytic conditions, followed by successive treatment of the resulting mixture of glucose and erythritol with sodium borohydride and acetic anhydride-pyridine, to give acetates of sorbitol and erythritol whose areas are measured by GLC. Another is ^{13}C-NMR analysis of the isolichenan, which has been used widely by Yokota et al.[22]

)–α–D–Glc*p*–(1→3)–α–D–Glc*p*–(1→3)–α–D–Glc*p*–(1→4)–

STRUCTURE 2.

C. α-D-Glucans Containing (1 → 3) and (1 → 4) Linkages

In 1972, Takeda et al.[23] found three glucans in *Evernia prunastri*; of these, two were α-D-glucans. Hot aqueous extraction of the lichen, followed by cooling of the extract, gave a precipitate that was purified by repeated dissolution, freezing, and thawing. It contained (1 → 3) and (1 → 4) linkages in a 4:1 ratio (0.75% yield). The supernatant was treated with cetyltrimethylammonium hydroxide, precipitating a β-D-glucan leaving in solution an α-D-glucan with the above linkages in a 3:2 ratio (0.43% yield). In the same investigation, an α-D-glucan was isolated from *Acroscyphus sphaerophoroides* (0.16% yield) and shown to contain (1 → 6) in addition to (1 → 3) and (1 → 4) linkages.

Nishikawa et al.[24] isolated cold water-insoluble α-D-glucans from 4 species of the genus *Cladonia: C. crispata* (0.57% yield), *C. mitis, C. rangiferina* (0.46% yield), and *C. squamosa* (0.28% yield). These contained alternate (1 → 3) and (1 → 4) linkages similar to nigeran (Structure 3). Similar glucans were later found in *C. alpestris* (Reindeer moss; 0.12% yield) and *C. confusa* (0.3% yield) except that the latter had a linkage ratio of 53:47.[25]

–α–D–Glc*p*–(1→3)–α–D–Glc*p*–(1→4)–

STRUCTURE 3.

As mentioned above, ^{13}C-NMR spectroscopy is useful in elucidation of glucan structure. An α-D-glucan, obtained by Takeda et al.[26] from *Parmelia caperata* in 3% yield, was water insoluble and contained (1 → 3) and (1 → 4) linkages in a 1:1 ratio, a value that was confirmed by Yokota et al.[22] by ^{13}C-NMR spectroscopy. In the same study, an α-D-glucan from *Cetraria richardsonii* was found to have these linkages in a 3:2 ratio, and is similar to isolichenan. Results obtained on the α-D-glucan of *A. sphaerophoroides* using conventional chemical analysis[23] were confirmed, namely that the relation of (1 → 3) to (1 → 4) linkages was 2:3 and that (1 → 6) linkages were also present. An identical glucan (1.9% yield) was obtained from *Sphaerophorus globosus*. Two members of the family Stereocaulaceae, *Stereocaulon japonicum* and *Pilophoron acicularis* (6 and 5% yield, respectively), contained (1 → 3) and (1 → 4) linkages in a 2:1 ratio. In these studies,[22] there was some variance between data obtained by NMR spectroscopy and conventional methods (see below).

A number of species of the genus *Stereocaulon* have been examined and their constituent glucans analyzed. In 1971, Hauan and Kjølberg[27] extracted *S. paschale* with hot water and purified a water-soluble glucan (0.45% yield) by successive freezing and thawing, followed by fractionation on a column of DEAE-cellulose. The product was a α-D-glucan having (1 → 3) and (1 → 4) linkages in a ratio of 1:2.5. An α-D-glucan of *S. japonicum*, isolated by Yokota and Shibata[28] in 1978 in a similar way, was analyzed by the methylation-hydrolysis-GLC method and found to have the linkages in a proportion of 3:1. However, a Smith degradation of the glucan resulted in the formation of *O*-α-D-glucopyranosyl-(1 → 3)-*O*-α-D-glucopyranosyl-(1 → 2)-D-erythritol only, a product consistent with a linkage ratio of 2:1. This proportion was confirmed in the following year using the technique of ^{13}C-NMR spectroscopy (see Reference 22). In 1981, Takahashi et al.[29] found that the α-D-glucans of *S. japonicum, S. sorediiferum* (1.8% yield), and *S. exutum* (5.4% yield) had (1 → 3) and (1 → 4) linkages in a molar ratio of 3:1, but lacking ^{13}C-NMR and Smith degradation data, it seems likely that the 3 glucans are identical but have 2:1 linkage ratios. Baron et al.[30] obtained from *S. ramulosum*, by hot water extraction, an α-D-glucan in 4.5% yield which had, according to a combination of methylation-hydrolysis-GLC, ^{13}C-NMR spectroscopic, and periodate oxidation analyses, (1 → 3) and (1 → 4) linkages in a ratio of 1.6:1. A Smith degradation provided erythritol, 2-*O*-α-D-glucopyranosyl-D-erythritol, and *O*-α-D-glucopyranosyl-(1 → 3)-*O*-α-D-glucopyranosyl-(1 → 2)-D-erythritol, showing that the (1 → 3) and (1 → 4) linkages are distributed somewhat irregulary along the linear chain.

Recently, Iacomini et al.[31] extracted *Letharia vulpina* with hot water and on cooling a precipitate of α-D-glucan formed (2% yield). It was also isolated in 4% yield by a similar extraction process carried out on lichen material previously treated with cold DMSO, which solubilized β-D-glucan. Methylation-hydrolysis-GLC analysis showed that (1 → 3) and (1 → 4) linkages were present in a 1.2:1 ratio, and the formation of 2-*O*-α-D-glucopyranosyl-D-erythritol and *O*-α-D-glucopyranosyl-(1 → 3)-*O*-α-D-glucopyranosyl-(1 → 2)-D-erythritol on Smith degradation is consistent with a regular distribution of the linkages along the linear chain. Although its composition is close to isolichenan and the "nigerans" isolated from *Cladonia* spp., it resembles the latter by virtue of its insolubility in cold water.

Gorin and Iacomini[19] in 1984 investigated the α-D-glucan of *Ramalina usnea* and found that extraction with hot water followed by cooling furnished a precipitate isolated in 0.5% yield. It likely contained amylose, giving a blue color with iodine, but it was insufficient to be detected by ^{13}C-NMR spectroscopy. Treatment of the extract with Fehling solution removed a galactomannan as its insoluble copper complex and from the supernatant an identical α-D-glucan was isolated in 2% yield. Methylation and Smith degradation data showed that (1 → 3) and (1 → 4) linkages were present in a 76:24 ratio (Structure 4).

)–α–D–Glc*p*–(1→3)–α–D–Glc*p*–(1→3)–α–D–Glc*p*–(1→3)–α–D–Glc*p*–(1→4)–

STRUCTURE 4.

In 1954, a cold water-soluble glucan was isolated from *Roccella montagnei* and called isolichenan, although detailed analyses were not carried out.[32]

D. β-D-Glucans Containing (1 → 3) and (1 → 4) Linkages or (1 → 3) Linkages Only

In 1972, Takeda et al.[23] extracted *E. prunastri* with hot water and, following removal of an α-D-glucan on successive freezing and thawing, the supernatant was treated with cetyltrimethylammonium hydroxide precipitating a β-D-glucan (0.63% yield). This differed from lichenan by virtue of its solubility in cold water and its 3:1 ratio of (1 → 3) to (1 → 4) linkages.

In 1974,[24] a β-D-glucan was obtained from *Usnea rubescens* (11.3% yield) by extraction with hot water followed by cooling, which caused its precipitation. It resembled lichenan since the molar ratio of (1 → 3) to (1 → 4) linkages was 3:7. Recently,[31] on subjecting an *Usnea* sp. to a similar procedure, a β-D-glucan was isolated but was found to be contaminated with traces of an α-D-glucan when examined by ^{13}C-NMR spectroscopy. However, the glucan could be isolated pure in 12% yield by extraction with cold DMSO followed by precipitation with ethanol. Methylation-hydrolysis-GLC-MS analysis indicated the presence of (1 → 3) and (1 → 4) linkages in a 1:3 ratio and on Smith degradation only erythritol and 2-*O*-β-D-glucopyranosyl-D-erythritol were detected. This showed that the linkage types are distributed regularly along the linear chain as in Structure 5. As these two glucans of *Usnea* spp. are different, there is some doubt as to the structure of "lichenan" isolated from *U. barbata* and advertized in the 1984 catalog of Sigma.

–)–β–D–Glc*p*–(1→3)–β–D–Glc*p*–(1→4)–β–D–Glc*p*–(1→4)–β–D–Glc*p*–(1→4)–

STRUCTURE 5.

In an early work, *U. longissima* was found to have a glucan with solubility properties corresponding to lichenan, but detailed structural analysis was not carried out.[32]

^{13}C-NMR spectroscopy was used to show that the β-D-glucan of *C. richardsonii* is lichenan.[22] The same study demonstrated that this tool is very useful in determining the purity of α- and β-D-glucan preparations as the signals of each group are readily distinguishable.

Cold DMSO extraction proved to be useful in preparation of a β-D-glucan from *L. vulpina* (12% yield) without contamination with α-D-glucan which is also present. Using analysis procedures identical to those applied to the *Usnea* sp. β-D-glucan described above, it was found to have Structure 5 also.[31]

A water-insoluble glucan, isolated from *S. ramulosum* in 0.19% yield, resembles laminaran in that it contains 3-*O*-substituted β-D-glucopyranosyl units only.[30] Its isolation in low yield and its structure typical of certain algae raises the question as to whether it arose from the phycobiont component.

E. β-D-Glucans Containing (1 → 6) Linkages

A glucan, isolated in 1943 by Drake[33] from *Umbilicaria pustulata* and called pustulin, was reinvestigated in 1954 by Lindberg and McPherson[34] (7% yield). It was soluble in hot water and precipitated on cooling, and was confirmed to be linear with 6-*O*-substituted β-D-glucopyranosyl residues. In 1968, Shibata et al.[35] found the glucan (presently called pustulan) in *Gyrophora esculenta* (5.2% yield[36]) and *Lasallia papulosa* (32% yield[36]). The IR spectra had bands corresponding to carbonyl absorption, whereas similar glucans of *U. pustulata* and *U. hirsuta* did not. In 1969, it was found that in glucans of *G. esculenta* and *L. papulosa*, the linkages were β-D-(1 → 6) and that approximately 10% of the units were substituted by *O*-acetyl groups.[37] Acetic acid, liberated on hydrolysis, was characterized by GLC and by formation of its *p*-bromophenacyl ester. The *O*-acetyl groups were found to be located at *O*-3 using the chemical method of Bouveng.[38] In 1970, Nishikawa et al.[39] investigated 3 species of the genus *Umbilicaria* — *U. angulata, U. caroliniana,* and *U. polyphylla* — which were found to contain similar *O*-acetylated glucans (7.2, 16, and 28% yield, respectively). From a taxonomic point of view, partially *O*-acetylated (1 → 6)-linked β-D-glucans are present in lichens of the same family Gyrophoraceae. This suggestion was confirmed to some extent by the isolation of similar *O*-acetylated polysaccharides from *Lasallia pensylvanica*[24] and *Actinogyra muehlenbergii*[31] (19.5 and 35% yield, respectively), but as already mentioned, some unacetylated glucans are known and, more significantly, a (1 → 6)-linked β-D-glucan has been found (4% yield) in an unrelated species, *Cladonia amaurocraea*.[25] Since it was extracted under alkaline conditions, it is not known whether it is acetylated in its native state.

The partly *O*-acetylated glucan of *A. muehlenbergii*, in order to avoid possible acetyl migration with hot water extraction prior to precipitation in the cold, was extracted with cold DMSO. ^{13}C-NMR spectral examination showed 6 signals of a (1 → 6)-linked β-D-glucopyranan and 2 minor peaks at δ 78.4 and 75.4 ppm corresponding to 3-*O*-acetylation, but the presence of another minor peak at δ 73.0 ppm indicated *O*-acetylation in another position[31] (Figure 2). The role of the glucan in this lichen is not clear, although it may help to conserve water in the rigorous conditions encountered on the surface of rocks. It appears to have some nutritive value, being called rock tripe (in Cree, Asīnīwākonak) and forms an edible soup on boiling with fish. Perhaps its value as a food is exaggerated, but it was eaten by the American revolutionary army in difficult times at Valley Forge. Another species — *G. esculenta* — is a food in Japan.

F. Branched β-D-Glucans Containing (1 → 3) and (1 → 6) Linkages

The glucan of *Cora pavonia*, which unlike the other described lichens has a basidiomycetous rather than a ascomycetous mycobiont, can be isolated in 1 to 2% yield by hot aqueous or aqueous alkali extraction followed by treatment with Fehling solution to remove a xylomannan as its insoluble Cu complex. It has the β-D-configuration and is highly branched with nonreducing end groups (21%), 3-*O*- (30%) and 6-*O*-substituted (30%) residues, and 3,6-di-*O*-substituted (19%) branch points. The main chain consists of interspersed (1 → 3) and (1 → 6) linkages.[40] The glucan is homogeneous on ultracentrifugation, electrophoresis, and gel filtration. Its branched structure is typical of β-D-glucans of basidiomycetes rather than α- and β-D-glucans of ascomycetes which are relatively linear having less than 10% of branches.[10]

G. Mannose-Containing Polysaccharides and Glycopeptides of Lichens

As early as 1906, *C. islandica*, was recognized by Ulander and Tollens[41] as containing galactose and mannose in its polysaccharide fraction. In 1933, Buston and Chambers[1] detected galacturonic acid also, and although this claim has not been substantiated since, the term "hemicellulose" was coined and unfortunately was accepted even until quite recently.[2] Karrer and Joos[42] in 1924 found that hot water extraction followed by cooling and removal of insoluble lichenan gave a supernatant which, on treatment with Fehling

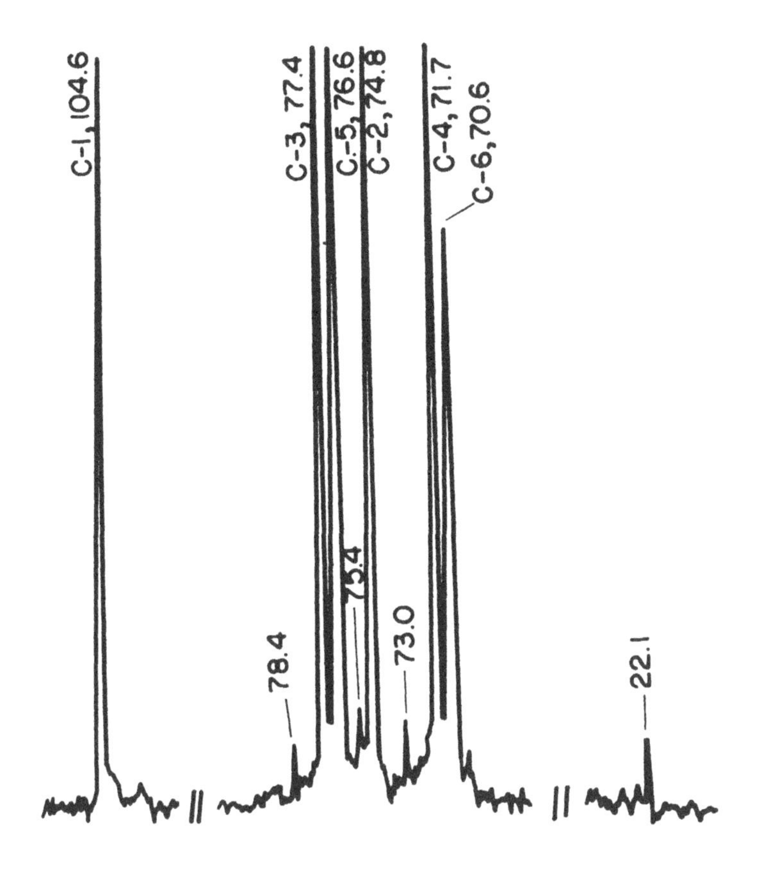

FIGURE 2. ^{13}C-NMR spectrum of partly *O*-acetylated glucan of *Actinogyra muehlenbergii* in deuterium oxide at 70° C. Numerical values are δ in parts per million.

solution, gave an insoluble Cu complex. This was regenerated to give a polymer containing mannose, galactose, and glucose in a molar ratio of 21:35:44 corresponding to a mixture of galactomannan and lichenan. Much later, Granichstädten and Percival[43] also obtained material consistent with a mixture of glucan and galactomannan. In another early study, Aspinall et al.[44] obtained from *C. alpestris* a Fehling precipitate containing the same monosaccharides.

Mićović et al.[45] in 1969 found in *E. prunastri* a heteropolysaccharide (3.5% yield) containing galactose, mannose, and galacturonic acid. It contained nonreducing end groups of galactopyranose (36%) and mannopyranose (11%) and tri-*O*-substituted mannopyranose residues, which by analogy with structures described below, may be part of a (1 → 6)-linked α-D-mannopyranosyl main chain.

Although mannose and galactose have been found frequently in lichen polysaccharides, there have been relatively few reports of galactomannans and of structural determinations carried out on them. One interesting observation, however, was made by Takahashi et al.,[3] who found a polymer (2.3% yield) containing mannose and galactose (3:2 ratio) in the culture-grown mycobiont of *C. mitis*.

Structural investigations on lichen galactomannans were mostly carried out recently by Iacomini, Gorin, and co-workers. The first, in 1984,[19] concerned *C. islandica* and *R. usnea*. The procedure of Karrer and Joos[42] was applied to prepare the galactomannan of the former. In *R. usnea*, a galactomannan and a water-soluble α-D-glucan were present and it was also necessary to treat the hot water extract with Fehling solution to selectively precipitate the galactomannan as its insoluble Cu complex. The structures of the 2 D-galacto-D-mannans (0.42 and 2.1% yield, respectively) were different, as indicated by C-1 portions of their

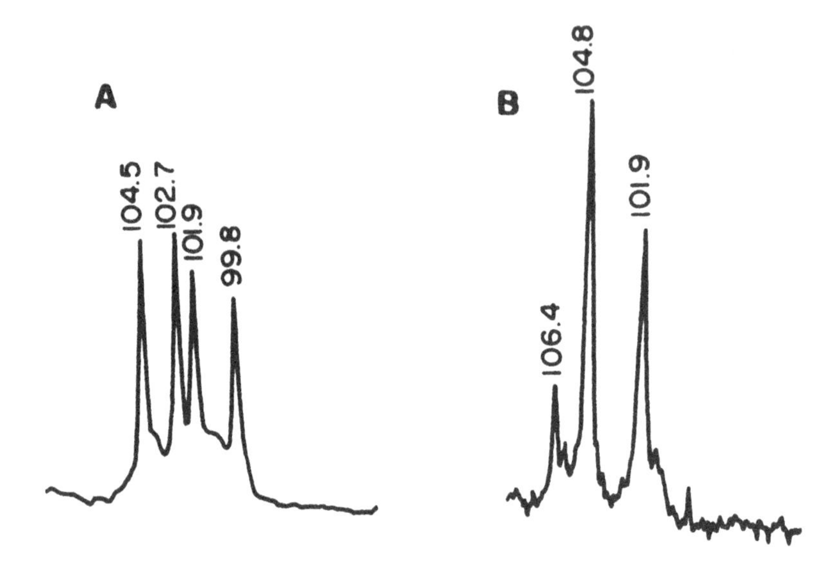

FIGURE 3. C-1 portions of ¹³C-NMR spectra of galactomannans in deuterium oxide at 70° C (numerical values are δ in parts per million): (A) of *Cetraria islandica*; (B) of *Ramalina usnea*.

¹³C-NMR spectra (Figure 3), and chemical analysis showed that while they both contained (1 → 6)-linked α-D-mannopyranosyl main chains, the patterns of side chain substitution were different. The structure of the galactomannan of *C. islandica* is more highly branched with the unusual feature, in the field of polysaccharide chemistry, of having two substituents on the same main chain residue (Structure 6); that of the *R. usnea* galactomannan has predominant Structure 7.

```
           α–D–Galp
               1
               ↓
               2
–α–D–Manp–(1→6)–α–D–Manp–(1→6)–
               4
               ↑
               1
           β–D–Galp
```

STRUCTURE 6.

```
–α–D–Manp–(1→6)–α–D–Manp–(1→6)–α–D–Manp–(1→6)–
      4                  4
      ↑                  ↑
      1                  1
  β–D–Galp           β–D–Galp
```

STRUCTURE 7.

In a comparative study on the galactomannans of *Cladonia* spp.,[25] 2 species with almost indistinguishable growth forms — *C. alpestris* (Central Saskatchewan, Canada) and *C. confusa* (coast of Southern Brazil) — were isolated by hot aqueous alkali extraction followed by precipitation with Fehling solution (3.2 and 1.6% yield, respectively). Their ^{13}C-NMR spectra were different from each other and in turn from those of *C. islandica* and *R. usnea* galactomannans. However, similarities occurred in the form of Structures 8 and 9. From the supernatants of the Fehling precipitations, structurally different polysaccharide components of *C. alpestris* and *C. confusa* were isolated (1.1 and 0.75% yield, respectively). The former contained a high proportion of Structure 10, in which a chain of (1 → 2)-linked α-D-mannopyranosyl units are substituted at *O*-6 by β-D-galactofuranosyl units. The structure of the D-galactan of *C. confusa* was relatively simple, since its ^{13}C-NMR spectrum contained 2 main C-1 signals at δ 102.2 and 108.8 ppm. Since the latter corresponds to β-D-galactofuranosyl units and since the specific rotation of the galactan is −47°, it bears a resemblance to the β-D-galactofuranan (−85°) of the phycobiont component of *R. crassa*. Such a furanosyl structure appears relatively common in phycobionts.[3] The other *Cladonia* sp., *C. amaurocraea*, gave rise to a Fehling precipitable galactomannan (5.6% yield) with the same components as those of the other 2 species, but its ^{13}C-NMR spectrum differed from those of all other galactomannans indicating fine structural differences.

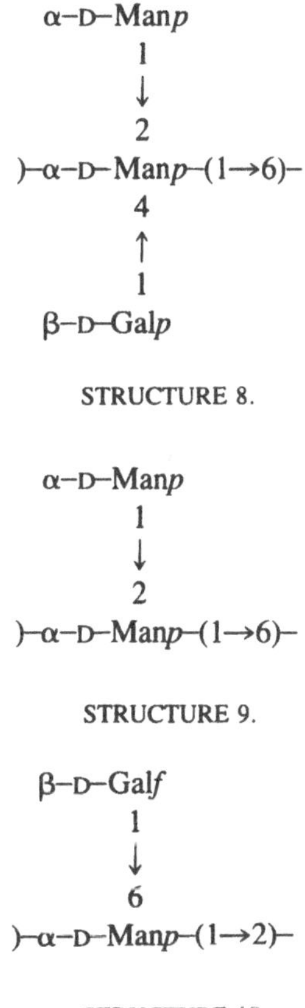

STRUCTURE 8.

STRUCTURE 9.

STRUCTURE 10.

In view of the structural diversity encountered in the Fehling-precipitated galactomannans of five lichens, demonstrated by their ^{13}C-NMR spectra, similar components of six more lichens were investigated by Gorin and Iacomini.[46] These were *Peltigera aphthosa, Parmelia sulcata, Letharia vulpina, Actinogyra muehlenbergii*, and an *Usnea* sp., all of which were collected in Canada. According to the ^{13}C-NMR spectra of galactomannans, most of which were isolated via their Fehling precipitates in 2 to 5% yield, it was confirmed that the structure of the galactomannan is typical of the lichen. Chemical analytical studies showed that in each case the galactomannan contained a (1 → 6)-linked α-D-mannopyranosyl main chain and that structural differences arose from the degree and sequence in which these residues are unsubstituted, or substituted at *O*-2, *O*-4, or *O*-2,4 by various side chains. Predominant structural components of the following lichens are *P. aphthosa* [Structure 11 (20%); Structure 12 (48%); and Structure 13 (6%)], *P. sulcata* (Structure 6), *Stereocaulon paschale* (Structures 8 and 12), and *A. muehlenbergii* (Structures 9 and 14). Thus, in the latter, β-D-galactofuranosyl residues were once again encountered in a lichen polysaccharide (0.6% yield).

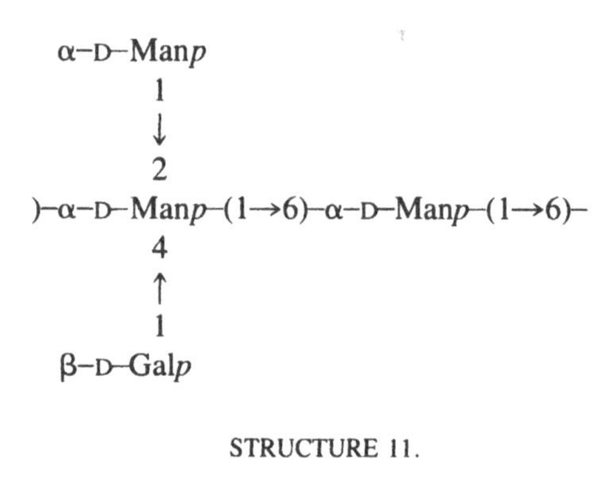

STRUCTURE 11.

)–α–D–Man*p*–(1→6)–
4
↑
1
β–D–Gal*p*

STRUCTURE 12.

)–α–D–Man*p*–(1→6)–

STRUCTURE 13.

β–D–Gal*p*
1
↓
4
)–α–D–Man*p*–(1→6)–
2
↑
1
α–D–Man*p*

STRUCTURE 14.

The mannans and galactomannans of ascomycetes are parts of glycopeptide complexes in their native state and can be liberated on treatment with aqueous alkali.[9] By analogy, similar complexes should be present in mycobionts of lichens. The first study on glycopeptides of lichens was carried out by Takahashi et al.[47] in 1974 who found them in cold aqueous extracts of Stictaceae. *Lobaria orientalis* contained at least two glycopeptides, but a study on one of them showed it to contain glucose and galactose with small quantities of mannose, arabinose, xylose, rhamnose, and glucosamine. The glucose units were possibly (1 → 6) linked, while those of mannose were (1 → 3) linked. The carbohydrate portion was connected to the serine and threonine portion of the peptide by *O*-glycosyl links. Also mentioned were the glycopeptides of lichens of the family Cladoniaceae, *C. belidiflora* and *C. graciliformis*, along with *A. sphaerophoroides*.

Although no attempts have been made so far in our laboratory to isolate glycopeptides from ascomycetous lichens, one was isolated from *Cora pavonia*, which has a basidiomycetous mycobiont.[40] Hot aqueous extraction followed by treatment with Fehling solution provided an insoluble Cu complex in very low yield. On regeneration, the complex was found to contain rhamnose, fucose, xylose, mannose, glucose, and galactose and had a protein content of 5.1%. It was homogeneous on ultracentrifugation, and on passage throughout a column of Sepharose gel 4B-200, and electrophoresis in aqueous barbital, pH 8.6, gave a band that was carbohydrate and protein (weakly) positive. The heteropolysaccharide portion contained mainly mannose and xylose, with a mannose-containing nucleus and a main chain with predominant (1 → 3)-linked α-D-mannopyranosyl units. These were unsubstituted [Structure 15 (10%)] and 4-*O*-[Structure 16 (10%)] and 2,4-di-*O*-substituted [Structure 17 (10%)] with residues of β-D-xylopyranose, thus representing 60% of total carbohydrate present. Such a structure is typical of cell walls of basidio- rather than ascomycetes,[10] whose mannose-containing polysaccharides have (1 → 6)-linked α-D-mannopyranosyl main chains. It bears a close resemblance to that of *Cryptococcus bacillosporus (neoformans)* serotype B. Sixty percent of the capsular polysaccharide consists of a (1 → 3)-linked α-D-mannopyranosyl main chain, one of every three units being substituted at *O*-2 by β-D-xylopyranosyl side-chains, and the other two being disubstituted at *O*-4 with β-D-xylopyranosyl and *O*-2 with β-D-glucopyranosyluronic acid units.[48]

−)−α−D−Man*p*−(1→3)−

STRUCTURE 15.

β−D−Xyl*p*
1
↓
4
)−α−D−Man*p*−(1→3)−

STRUCTURE 16.

β−D−Xyl*p*
1
↓
4
)−α−D−Man*p*−(1→3)−
2
↑
1
β−D−Xyl*p*

STRUCTURE 17.

III. LOW MOLECULAR WEIGHT CARBOHYDRATES IN LICHENS

In green phycobionts, the carbohydrates produced by photosynthesis are polyols, namely sorbitol in *Hyalococcus,* ribitol in *Coccomyxa, Myrmecia,* and *Trebouxia*, and erythritol in *Trentepohlia*. The photosynthetic carbohydrate produced by the cyanobiont *Nostoc* is glucose.[49,50] Sucrose is also formed, along with ribitol, when *Trebouxia* is freshly isolated from the thallus[51] (see also Volume I, Chapters VI.A and VI.C).

Phycobiont precursors are converted into mannitol[52] and, if the lichen continues photosynthesis, it is transposed into starch grains in the algal chloroplasts and, to a lesser extent, in grains of the fungus.[53] The biosynthesis of arabinitol has also been studied.[54] Carbohydrate transferred by the phycobiont eventually ends up in the mycobiont in the form of lichenan, isolichenan, and protein.[55]

D-mannitol is the most common polyol in lichens; it was found in all 80 species examined,[56] compared with D-arabinitol, which was found in 55 of 60 lichens examined.[57] In GLC quantitation studies, arabinitol was found to predominate in hot aqueous methanol extracts. In 10 lichens examined, mannitol and arabinitol were found, respectively, in the following amounts (dry weight): *C. alpestris* (0.3 and 1.4%), *C. confusa* (0.4 and 2.2%), *C. amaurocraea* (0.4 and 1.5%),[25] *A. muehlenbergii* (0.6 and 1.2%), *P. aphthosa* (2.4 and 2.8%), *Usnea* sp. (0.4 and 4.0%), *L. vulpina* (0.4 and 2.5%), *P. sulcata* (0.3 and 2.5%), *S. paschale* (0.8 and 2.5%),[46] and *R. usnea* (<0.0 and 1.0%).[19]

Other polyols encountered in lichens are glycerol, *myo*-inositol, xylitol, siphulitol (1-deoxy-D-*glycero*-D-*talo*-heptitol), and volemitol (D-*glycero*-D-*talo*-heptitol). Oligosaccharides include trehalose, sucrose, 3-*O*-β-D-glucopyranosyl-D-mannitol, peltigeroside (3-*O*-β-D-galactofuranosyl-D-mannitol), umbilicin (2-*O*-β-D-galactofuranosyl-D-arabinitol), 1-*O*-β-D-galactopyranosyl-D-ribitol, *O*-β-D-galactopyranosyl (1 → 6)-*O*-β-D-galactopyranosyl-(1 → 1)-D-glycerol, a mannosido-mannitol, and a galactosyl pentitol. Monosaccharides reported are arabinose, fructose, galactose, tagatose, xylose, lyxose, ribose, mannose, rhamnose, and glucosamine, although some of the characterizations may be doubtful. The functions of these components in lichens is not yet well defined. Because of the large number of investigations that have been carried out on identification of these materials, they are not reported individually herein. The reader is referred to the excellent chemical guides to lichen products prepared by Chicita Culberson et al.[58-60] and to the reviews of Pueyo.[61-65]

Complete GLC analysis of low molecular weight lichen components, including disaccharides, can be carried out using acetyl, trifluoroacetyl, and trimethylsilyl derivatives. In addition to polyols, trehalose, 3-*O*-β-D-glucopyranosyl-D-mannitol, and 2-*O*-β-D-galactofuranosyl-D-arabinitol may be detected.[66]

Cora pavonia is unusual in that it contains 4.4% of α-α-D-trehalose which is insoluble on extraction, being contamined with only 0.13% of mannitol.[40] This lichen may prove to be a valuable source of the disaccharide.

IV. AMINO ACID AND PROTEIN COMPONENTS OF LICHENS

The free and bound amino acids of lichens are of interest in evaluating the potential of lichens in nutrition. The best-known edible lichen is Reindeer moss (*Cladonia alpestris*), which is consumed by the European reindeer and its North American counterpart, the caribou. However, the preferred food of these animals is grasses; browsing on lichens is a last resort. It was found that reindeer calves, fed over a period of 5 winter months on lichens, developed severe malnutrition[67] (see also Chapter XII.B).

The nitrogen content of the thallus can vary from 0.33% in *C. impexa* to 9.2% in *Lecanora muralis*. Nitrogen content of 28 lichen species is summarized in Table 1 in Chapter VI.B of Volume I. Although it is not a complete list, it serves to show the range of possible values. Those of *Cladonia* spp. are at the lower end of the scale. Generally, lichens with cyanobionts have higher nitrogen contents than those with green phycobionts, except when the latter grow on nitrogen-rich substrates.[69] In *Cora pavonia*, which contains a cyanobiont, amino acids and protein represent 36% of total lichen, and one tenth of this is tyrosine.[40]

In lichens, amino acids are present in addition to proteins, and in *Sticta sylvatica, Lobaria laetevirens,* and *L. pulmonaria*, glutamic acid is the most abundant[70] (see also Tables 3 and 4 in Chapter VI.B of Volume I). Although it might be imagined that other components contribute to total nitrogen, it is pertinent that in total amino acid determinations following hydrolysis, recoveries of 73 to 88% are made, based on nitrogen content.[25,46] Free amino acid composition can vary according to the season.[71] A number of qualitative and/or quantitative analyses have been carried out on free amino acids[72-77] and on free amino acid and protein[71,76] contents of lichens. In one study, 28 lichens were examined and found to contain 11 to 17 different free amino acids.[76] Including 75 lichens examined previously,[75] the predominant components were found to be alanine and glutamic acid; 49 species contained >100 μmol of alanine and 20 contained >100 μmol of glutamic acid. Taurine was a significant component.

REFERENCES

1. **Buston, H. W. and Chambers, V. H.,** Some cell-wall constituents of *Cetraria islandica* ("Iceland moss"), *Biochem. J.*, 27, 1691, 1933.
2. **Ozenda, P. and Clauzade, G.,** *Les Lichens,* Masson et Cie, Paris, 1970, 71.
3. **Takahashi, K., Takeda, T., and Shibata, S.,** Polysaccharides of lichen symbionts, *Chem. Pharm. Bull.*, 27, 238, 1979.
4. **Brown, R. M. and Wilson, R.,** Electron microscopy of the lichen *Physcia aipolia* (Ehrh.) Nyl., *J. Phycol.*, 4, 230, 1968.
5. **Jacobs, J. B. and Ahmadjian, V.,** The ultrastructure of lichens. I. A general survey, *J. Phycol.*, 5, 227, 1969.
6. **Ahmadjian, V.,** Lichens, in *Symbiosis,* Vol. 1, Henry, M., Ed., Academic Press, New York, 1966, 35.
7. **Hough, L., Jones, J. K. N., and Wadman, W. H.,** An investigation of the polysaccharide components of certain fresh-water algae, *J. Chem. Soc.*, 3393, 1952.
8. **Chao, L. and Bowen, C. C.,** Purification and properties of glycogen isolated from a blue-green alga, *Nostoc muscorum, J. Bacteriol.*, 105, 331, 1971.
9. **Gorin, P. A. J. and LaRue, T. M.,** unpublished data.
10. **Gorin, P. A. J. and Barretto-Bergter, E.,** The chemistry of polysaccharides of fungi and lichens, in *The Polysaccharides,* Vol. 2, Aspinall, G. O., Ed., Academic Press, New York, 1983, 366.
11. **Berzelius, J. J.,** Versuche über die Mischung des Isländischen Mooses und seine Anwendung als Nahrungsmittel, *J. Chim. Phys.*, 7, 317, 1815.
12. **Varry, G.,** Ueber zwei als Gummiarten Betrachtete Natürliche Pflanzenprodukte, *Ann. Chim.*, 13, 71, 1835
13. **Meyer, K. H. and Gürtler, P.,** Recherches sur l'amidon. XXXI. La constitution de la lichénine, *Helv. Chim. Acta,* 30, 751, 1947.
14. **Chanda, N. B., Hirst, E. L., and Manners, D. J.,** A comparison of lichenin and isolichenin from Iceland moss (*Cetraria islandica*), *J. Chem. Soc.*, 1951, 1957.
15. **Peat, S., Whelan, W. J., and Roberts, J. G.,** The structure of lichenin, *J. Chem. Soc.*, 3916, 1957.
16. **Perlin, A. S. and Suzuki, S.,** The structure of lichenin: selective enzymolysis studies, *Can. J. Chem.*, 40, 50, 1962.
17. **Fleming, M. and Manners, D. J.,** A comparison of the fine-structure of lichenin and barley glucan, *Biochem. J.*, 100, 4p, 1966.
18. **Meyer, K. H. and Gürtler, P.,** Recherches sur l'amidon. XXXII. L'isolichénine, *Helv. Chim. Acta,* 30, 761, 1947.
19. **Gorin, P. A. J. and Iacomini, M.,** Polysaccharides of the lichens *Cetraria islandica* and *Ramalina usnea, Carbohydr. Res.*, 128, 119, 1984.
20. **Peat, S., Whelan, W. J., Turvey, J. R., and Morgan, K.,** The structure of isolichenin, *J. Chem. Soc.*, 623, 1961.
21. **Fleming, M. and Manners, D. J.,** The fine structure of isolichenin, *Biochem. J.*, 100, 24p, 1966.
22. **Yokota, I., Shibata, S., and Saitô, H.,** A ^{13}C-NMR analysis of linkages in lichen polysaccharides: an approach to chemical taxonomy of lichens, *Carbohydr. Res.*, 69, 252, 1979.

23. **Takeda, T., Funatsu, M., Shibata, S., and Fukuoka, F.,** Polysaccharides of lichens and fungi. V. Antitumor active polysaccharides of lichens of *Evernia, Acroscyphus, Alectoria* spp., *Chem. Pharm. Bull.*, 20, 2445, 1972.
24. **Nishikawa, Y., Ohki, K., Takahashi, K., Kurono, G., Fukuoka, F., and Emori, M.,** Studies on the water-soluble constituents of lichens. II. Antitumor polysaccharides of *Lasallia, Usnea,* and *Cladonia* spp., *Chem. Pharm. Bull.*, 22, 2692, 1974.
25. **Iacomini, M., Schneider, C. L., and Gorin, P. A. J.,** Comparative studies on the polysaccharides of *Cladonia alpestris* (Reindeer moss), *Cladonia confusa,* and *Cladonia amaurocraea, Carbohydr. Res.*, 142, 237, 1985.
26. **Takeda, T., Nishikawa, Y., and Shibata, S.,** A new α-glucan from the lichen *Parmelia caperata* (L.) Ach., *Chem. Pharm. Bull.*, 18, 1074, 1970.
27. **Hauan, E. and Kjølberg, O.,** Studies on the polysaccharides of lichens. I. The structure of a water-soluble polysaccharide in *Stereocaulon paschale* (L.) Fr., *Acta Chem. Scand.*, 25, 2622, 1971.
28. **Yokota, I. and Shibata, S.,** A polysaccharide of the lichen, *Stereocaulon japonicum, Chem. Pharm. Bull.*, 26, 2668, 1978.
29. **Takahashi, K., Kon, T., Yokota, I., and Shibata, S.,** Chemotaxonomic studies on the polysaccharides of lichens. Polysaccharides of stereocaulaceous lichens, *Carbohydr. Res.*, 89, 166, 1981.
30. **Baron, M., Gorin, P. A. J., and Iacomini, M.,** unpublished results.
31. **Iacomini, M., Gorin, P. A. J., Baron, M., Tulloch, A. P., and Mazurek, M.,** Novel glucans obtained on dimethyl sulfoxide extraction of the lichens, *Letharia vulpina, Actinogyra muehlenbergii,* and an *Usnea* sp., *Carbohydr. Res.*, submitted.
32. **Mittal, O. P. and Seshadri, T. R.,** Chemical investigation of Indian lichens. XVI. Purification and composition of lichenin and isolichenin from Indian lichens, *J. Sci. Ind. Res. (India)*, 13B, 244, 1954.
33. **Drake, B.,** Some polyglucides of lichens, particularly lichenin and newly discovered pustulin, *Biochem. Z.*, 313, 388, 1943.
34. **Lindberg, B. and McPherson, J.,** Studies on the chemistry of lichens. VI. The structure of pustulan, *Acta Chem. Scand.*, 8, 985, 1954.
35. **Shibata, S., Nishikawa, Y., Takeda, T., Tanaka, M., Fukuoka, F., and Nakanishi, M.,** Studies on the chemical structures of new glucans isolated from *Gyrophora esculenta* Miyoshi and *Lasallia papulosa* (Ach.) Llano and their inhibiting effect on implanted sarcoma-180 in mice, *Chem. Pharm. Bull.*, 16, 1639, 1968.
36. **Shibata, S., Nishikawa, Y., Takeda, T., and Tanaka, M.,** Polysaccharides in lichens and fungi. I. Antitumor active polysaccharides of *Gyrophora esculenta* Miyoshi and *Lasallia papulosa* (Ach.) Llano, *Chem. Pharm. Bull.*, 16, 2362, 1968.
37. **Nishikawa, Y., Takeda, T., Shibata, S., and Fukuoka, F.,** Polysaccharides in lichens and fungi. III. Further investigation on the structures and the antitumor activity of the polysaccharides from *Gyrophora esculenta* Miyoshi and *Lasallia papulosa* (Ach.) Llano, *Chem. Pharm. Bull.*, 17, 1910, 1969.
38. **Bouveng, H. O.,** Phenylisocyanate derivatives of carbohydrates, *Acta Chem. Scand.*, 15, 87, 1961.
39. **Nishikawa, Y., Tanaka, M., Shibata, S., and Fukuoka, F.,** Polysaccharides of lichens and fungi. IV. Antitumor active *O*-acetylated pustulan-type glucans from the lichens of *Umbilicaria* species, *Chem. Pharm. Bull.*, 18, 1431, 1970.
40. **Iacomini, M., Zanin, S. M. W., Fontana, J. D., and Gorin, P. A. J.,** Isolation and characterization of β-D-glucan, heteropolysaccharides and trehalose components of the basidiomycetous lichen *Cora pavonia, Carbohydr. Res.*, 168, 55, 1987.
41. **Ulander, A. and Tollens, B.,** Untersuchungen über die Kohlenhydrate der Flechten, *Chem. Ber.*, 39, 401, 1906.
42. **Karrer, P. and Joos, B.,** Polysaccharide. XXX. Zur Kenntnis des Isolichenins, *Hoppe-Seyler's Z. Physiol. Chim.*, 141, 311, 1924.
43. **Granichstädten, H. and Percival, E. G. V.,** The polysaccharides of Iceland moss (*Cetraria islandica*). I. Preliminary study of the hemicelluloses, *J. Chem. Soc.*, 54, 1943.
44. **Aspinall, G. O., Hirst, E. L., and Warburton, M.,** The alkali-soluble polysaccharides of the lichen *Cladonia alpestris* (Reindeer moss), *J. Chem. Soc.*, 651, 1955.
45. **Mićović, V. M., Hranisavljević-Jakovljević, M., and Miljković-Stojanović, J.,** Structural study of polysaccharides from the oak lichen *Evernia prunastri* (L.) Ach., *Carbohydr. Res.*, 10, 525, 1969.
46. **Gorin, P. A. J. and Iacomini, M.,** Structural diversity of D-galacto-D-mannan components isolated from lichens having ascomycetous mycosymbionts, *Carbohydr. Res.*, 142, 253, 1985.
47. **Takahashi, K., Takeda, T., Shibata, S., Inomata, M., and Fukuoka, F.,** Polysaccharides of lichens and fungi. VI. Antitumor active polysaccharides of lichens of Stictaceae, *Chem. Pharm. Bull.*, 22, 404, 1974.
48. **Bhattacharjee, A. K., Kwon-Chung, K. J., and Glaudemans, C. P. J.,** On the structure of the capsular polysaccharide from *Cryptococcus neoformans* serotype C. II., *Mol. Immunol.*, 16, 531, 1979.

49. **Richardson, D. H. S., Hill, D. H., and Smith, D. C.,** Lichen physiology. XI. The role of the alga in determining the pattern of carbohydrate movement between lichen symbionts, *New Phytol.*, 67, 469, 1968.
50. **Hill, D. J.,** The Carbohydrate Movement between the Symbionts of Lichens, D. Phil. thesis, University of Oxford, 1970.
51. **Green, T. G. A.,** The Biology of Lichen Symbionts, D. Phil. thesis, University of Oxford, 1970.
52. **Drew, E. A. and Smith, D. C.,** Studies in the physiology of lichens. VIII. Movement of glucose from alga to fungus during photosynthesis in the thallus of *Peltigera polydactyla*, *New Phytol.*, 66, 389, 1967.
53. **Jacobs, J. B. and Ahmadjian, V.,** The ultrastructure of lichens. VI. Movement of carbon products from alga to fungus as demonstrated by high resolution radioautography, *New Phytol.*, 70, 47, 1971.
54. **Richardson, D. H. S. and Smith, D. C.,** Lichen physiology. IX. Carbohydrate movement from the *Trebouxia* symbiont of *Xanthoria aureola* to the fungus, *New Phytol.*, 67, 61, 1968.
55. **Hale, M. E.,** *The Biology of Lichens*, Arnold, London, 1967.
56. **Lewis, D. H. and Smith, D. C.,** Sugar alcohols (polyols) in fungi and green plants. I. Distribution, physiology, and metabolism, *New Phytol.*, 66, 143, 1967.
57. **Lindberg, B., Misiorny, A., and Wachtmeister, C. A.,** Studies on the chemistry of lichens. IV. Investigation of low molecular weight carbohydrate constituents of different lichens, *Acta Chem. Scand.*, 7, 591, 1953.
58. **Culberson, C. F.,** *Chemical and Botanical Guide to Lichen Products*, University of North Carolina Press, Chapel Hill, 1969.
59. **Culberson, C. F.,** First supplement to "Chemical and Botanical Guide to Lichen Products", *Bryologist*, 73, 177, 1970.
60. **Culberson, C. F., Culberson, W. L., and Johnson, A.,** Second supplement to "Chemical and Botanical Guide to Lichen Products", The American Bryological and Lichenological Society, St. Louis, Mo., 1977.
61. **Pueyo, G.,** Carbohydrate constituents in lichens. VI. Osides, *Ann. Falsif. Expert. Chim. Toxicol.*, 77, 67, 1984.
62. **Pueyo, G.,** Carbohydrate constituents in lichens. VII. Conclusion bibliographic review, *Ann. Falsif. Expert. Chim. Toxicol.*, 77, 203, 1984.
63. **Pueyo, G.,** Carbohydrate constituents of lichens. IV. Osides, *Ann. Falsif. Expert. Chim. Toxicol.*, 75, 123, 1982.
64. **Pueyo, G.,** Glucidic constituents of lichens. III. Osides, *Ann. Falsif. Expert. Chim. Toxicol.*, 74, 343, 1981.
65. **Pueyo, G.,** Study of lichen carbohydrates, *Bull. Cent. Étud. Rech. Sci. Biarritz*, 12, 209, 1978.
66. **Nishikawa, Y., Michishita, K., and Kurono, G.,** Water-soluble constituents of lichens. I. Gas chromatographic analysis of low molecular weight carbohydrates, *Chem. Pharm. Bull.*, 21, 1014, 1973.
67. **Bjarghov, R. S., Fjellheim, P., Hove, K., Jacobsen, E., Skjenneberg, S., and Try, K.,** Nutritional effects on serum enzymes and other blood constituents in reindeer calves *(Rangifer tarandus tarandus)*, *Comp. Biochem. Physiol.*, 55, 187, 1976.
68. **Hitch, C. J. B.,** A Study of Some Environmental Factors Affecting Nitrogenase Activity in Lichens, M.Sc. thesis, University of Dundee, Scotland, 1971.
69. **Massé, L. C.,** Étude comparée des teneurs en azote des lichens et de leurs substrate: les espèces ornithocoprophiles, *C. R. Acad. Sci. Ser. D*, 262, 1721, 1966.
70. **Goas, G. and Bernard, T.,** Contribution à létude du metabolisme azote des lichens: les différentes formes d'azote de quelques espèces de la famille des Stictacées, *C. R. Acad. Sci. Ser. D*, 265, 1187, 1967.
71. **Perseca, T., Codoreanu, V., [illegible], I., and Petrosian, F.,** Free and protein amino acids in different lichen species of Romania, *Stud. Univ. Babes-Bolyai Ser. Biol.*, 26, 27, 1981.
72. **Perseca, T., Dordea, M., and Codoreanu, V.,** Studies on free amino acid content in several lichen species, *Stud. Univ. Babes-Bolyai Ser. Biol.*, 24, 26, 1979.
73. **Margaris, N. S.,** Free amino acid pools in *Cladonia pyxidata* and *Peltigera* species, *Bryologist*, 77, 77, 1974.
74. **Fujikawa, F., Hirayama, T., Mori, T., Nakamura, M., Takagaki, Y., Motoda, Y., Mukainaka, H., Ito, K., Matsui, R., and Fujisawa, M.,** Free amino acids in lichens of Japan. III., *Yakugaku Zasshi*, 93, 1558, 1973.
75. **Fujikawa, F., Hirai, K., Hirayama, T., Toyota, T., Nakamura, T., Nishimaki, T., Yoshikawa, T., Yasuda, S., Nishio, S., Kojitani, K., Nakai, T., Ando, T., Tsuji, Y., Tomisaki, K., Watanabe, M., Fujisawa, M., Nagai, M., Koyama, M., Matsoami, N., Urasaki, M., and Takagawa, M.,** Free amino acids in Japanese lichens. I., *Yakugaku Zasshi*, 90, 1267, 1970.
76. **Fujikawa, F., Hirai, K., Hirayama, T., Toyota, T., Urasaki, M., Takagawa, M., Fukuda, M., Moritani, K., Tomoike, S., Harada, H., and Matoba, C.,** Free amino acids in Japanese lichens. II., *Yakugaku Zasshi*, 92, 823, 1972.
77. **Budriene, S. and Jankevicius, K.,** Amino acid composition of some species of lichens of Lithuanian SSR in comparison to the amino acid composition of lower plants, *Liet. TSR Mokslu Akad. Darb. Ser. C*, 3, 1983.

Chapter IX.C.1

CAROTENOIDS

Bazyli Czeczuga

Carotenoids are pigments varying in color from yellow to red which occur in cells of all auto- and heterotrophic plants. They are insoluble in water, but readily dissolve in organic solvents such as ethanol, methanol, hexane, ether, and chloroform. They are photolabile; i.e., they are easily destroyed in the presence of oxygen and light.[1,2]

From the chemical aspect, carotenoids are tetraterpenes built of eight isoprene units bound so that the two methyl groups nearest the molecule center are in position 1,6, whereas the other methyl groups are in positions 1 and 5. These pigments have a number of conjugated double and even triple bonds forming a chromatophore system with a large number of cis-trans isomers. Under natural conditions, carotenoids usually occur in the trans form. Carotenoids form a subgroup of polyene pigments generally containing 40 carbon atoms in the molecule (Figure 1). In some bacteria, carotenoids of C_{30}, C_{45}, and C_{50} occur. All the natural carotenoids can be regarded as having been derived from lycopene by reactions involving (1) hydrogenation, (2) dehydrogenation, (3) cyclization, (4) insertion of oxygen in various forms, (5) double bond migration, (6) methyl migration, (7) chain elongation, and (8) chain shortening.[3]

The biosynthesis of carotenoids in plants begins in a pathway common to all terpenes; the first specific carotenoid precursor in the pathway is phytoene. Beginning with acetyl-CoA, three reactions occur which require an influx of energy and phosphoric acid radicals from ATP and result in the formation of isopentenyl pyrophosphate (Figure 2). From isopentenyl pyrophosphate through a series of condensations mediated by specific isomerases, geranyl pyrophosphate, farnesyl pyrophosphate, and geranyl geranyl pyrophosphate are formed. The characteristic acyclic tetraterpene of carotenoids (e.g., phytoene) is formed through the dimerization of 20-C molecules of geranyl geranyl pyrophosphate (Figure 2). In addition, cyclic carotenoids may be formed from 40-C acyclic precursors of neurosporene or lycopene.

Generally speaking, the naturally occurring carotenoids can be divided into six groups. They include

1. Hydrocarbon carotenoids generally termed carotenes with an empirical formula of $C_{40}H_{56}$. The most common of these found in nature, α-, β-, and γ-carotenoids, are built of one or two α or β-ionone rings. The names of the various carotenoids are derived from the names of the specific end groups on the carotenoid. The end groups and their prefixes are given in Table 1 and structures of the end groups in Figure 3. Each carotenoid has a common name and a semisystematic name. With respect to carotenes, the semisystemic names of lycopenes are ψ,ψ-carotene, α-carotene, βε-carotene, β-carotene, β,β-carotene, γ-carotene, and β,ψ-carotene.
2. Oxygenated carotenoids generally termed xanthophylls with an empirical formula of $C_{40}H_{56}O_{1-8}$. Oxygen may become bound to the ionone rings in the form of a hydroxyl group (–OH) as in zeaxanthin, an epoxide group (–O–) as in mutatoxanthin, a methoxyl group (– COH_3) as in spirilloxanthin, a ketone group (–C=O) as in canthaxanthin, a carboxyl group (–COOH) as in torularhodin, an aldehyde (–CHO) as in torularhodin aldehyde, or a glycosyloxy group (–$C_6H_{11}O_5$) as in myxoxanthophyll. These functional groups determine the different physicochemical properties of xanthophylls.
3. Retrocarotenoids are carotenoids in which all the single and double bonds of conjugated

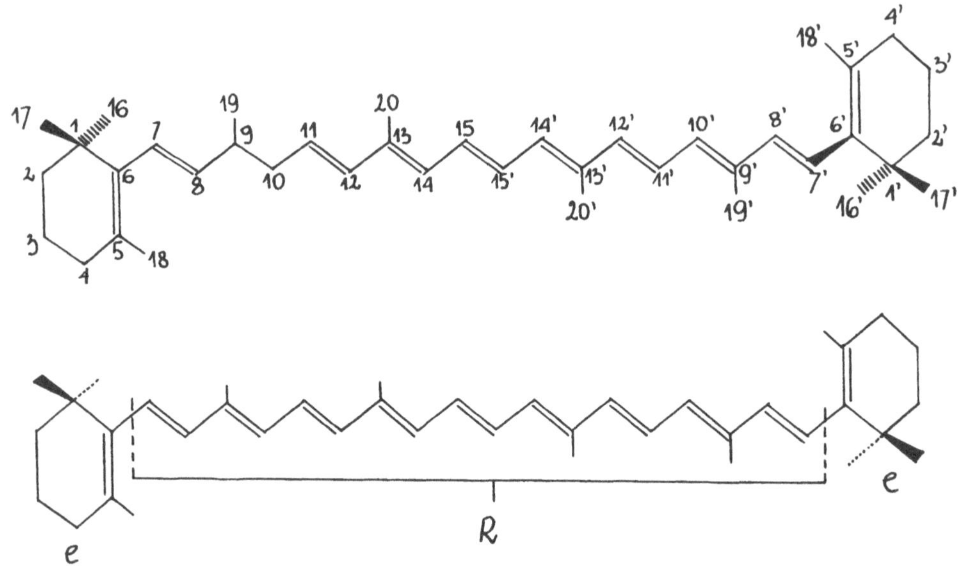

FIGURE 1. The structure of β-carotene and numbering of the carbon atoms. e, end group; R, bond.

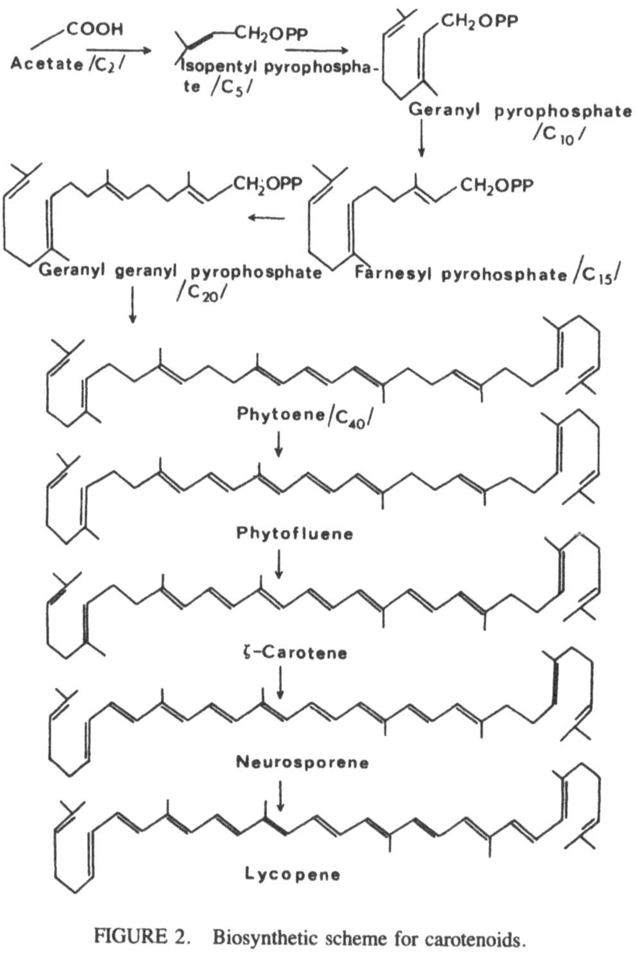

FIGURE 2. Biosynthetic scheme for carotenoids.

Table 1
END GROUP DESIGNATION OF CAROTENOIDS

Type	Prefix	Formula	Structure[a]	Example
Acyclic	ψ (psi)	C_9H_{15}	A	Lycopene
Cyclohexene	β,ε (beta, epsilon)	C_9H_{15}	B, C	α-Carotene
Methylenecyclohexene	γ (gamma)	C_9H_{15}	D	"β, γ-Carotene"
Cyclopentene	κ (kappa)	C_9H_{17}	E	Capsorubin
Aryl	φ, χ (phi, chi)	C_9H_{11}	F, G	Renieratene

[a] See Figure 3.

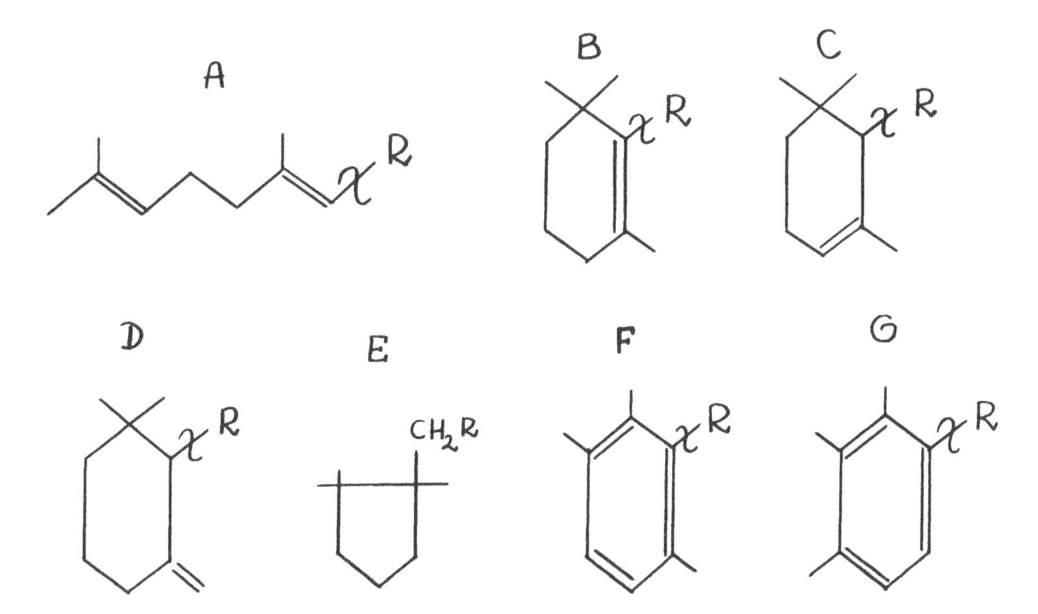

FIGURE 3. Structure of the end groups.

polyene systems are shifted by one place. Rhodoxanthin is an example from this group which is found in lichens.

4. Seco and apocarotenoids. Secocarotenoids are those in which the bond between two bound carbon atoms (different from the 1 and 6 end cyclic groups) has been broken. Apocarotenoids, on the other hand, are those in which the carbon skeleton has been shortened as a result of the removal of fragments at one or both ends of the molecule. The prefix "apo" is preceded by the number of the carbon atom beyond which the fragment of the molecule of the given carotenoid has been replaced by a hydrogen atom.
5. Nor-carotenoids are ones in which the CH_3, CH_2, or CH groups have been eliminated. Before the prefix "nor", the number of the eliminated carbon atom is given.
6. Higher carotenoids is the name given to the group of hydrocarbons or their oxygen derivatives in which the molecule is built of more than eight isoprenoid radicals (e.g., C_{45} or C_{50}) bound in a similar way to that of C_{40} carotenoids.

With respect to lichens, 143 species representing 13 families have so far been investigated.[4-9] From these, 42 carotenoids have been determined (Table 2 and Figure 4).

Table 2
LIST OF CAROTENOIDS FROM THE LICHEN SPECIES INVESTIGATED

No.	Carotenoid	Structure[a]	Semisystematic name	Group[b]	Occurrence[c]	
					a + c	f
	Most frequent					
1	β-Carotene	g − R − g	β,β-Carotene	hc	+	+
2	β-Cryptoxanthin	g − R − i	β,β-Caroten-3-ol	xh	+	+
3	Zeaxanthin	i − R − i	β,β-Carotene-3,3′-diol	xh	+	+
4	Lutein	i − R − k	β,ε-Carotene-3,3′-diol	xh	+	
5	Astaxanthin	o − R − o	3,3′-Dihydroxy-β,β-carotene-4,4′-dione	xk	+	+
6	Lutein epoxide	k − R − r	5,6-Epoxy-5,6-dihydro-β,ε-carotene-3,3′-diol	xe	+	
7	Mutatoxanthin	i − R_1 − t	5,8-Epoxy-5,8-dihydro-β,β-carotene-3,3′-diol	xe	+	
	Frequent					
8	Lycopene	a − R − a	ψ,ψ-Carotene	hc	+	+
9	α-Carotene	g − R − h	β,ε-Carotene	hc	+	
10	Lycoxanthin	a − R − d	ψ,ψ-Caroten-16-ol	xh		+
11	Lycophyll	d − R − d	ψ,ψ-Carotene-16,16′-diol	xh		+
12	α-Cryptoxanthin	h − R − i	β,ε-Caroten-3-ol	xh	+	
13	Echinenone	g − R − n	β,β-Caroten-4-one	xk	+	+
14	Canthaxanthin	n − R − n	β,β-Carotene-4,4′-dione	xk	+	+
15	Phoenicoxanthin	n − R − o	3-Hydroxy-β,β-carotene-4,4′-dione	xk	+	
16	Adonixanthin	i − R − o	3,3′-Dihydroxy-β,β-caroten-4-one	xk	+	
17	α-Doradexanthin	k − R − o	3,3′-Dihydroxy-β,ε-caroten-4-one	xk	+	
18	Diatoxanthin	i − R_1 − l	7,8-Didehydro-β,β-carotene-3,3′-diol	xh	+	
19	β-Carotene epoxide	g − R − p	5,6-Epoxy-5,6-dihydro-β,β-carotene	xe	+	
20	Antheraxanthin	i − R − r	5,6-Epoxy-dihydro-β,β-carotene-3,3′-diol	xe	+	
21	Neoxanthin	r − R_1 − m	5′,6′-Epoxy-6,7-didehydro-5,6,5′,6′-tetrahydro-β,β-carotene-3,5,3′-triol	xe	+	
22	Violaxanthin	r − R − r	5,6,5′,6′-Diepoxy-5,6,5′,6′-tetrahydro-β,β-carotene-3,3′-diol	xe	+	
23	Mutatochrome	g − R_1 − s	5,8-Epoxy-5,8-dihydto-β,β-carotene	xe	+	
24	Aurochrome	s − R_2 − s	5,8,5′,8′-Diepoxy-5,8,5′,8′-tetrahydro-β,β-carotene	xe	+	
25	Auroxanthin	t − R_2 − t	5,8,5′,8′-Diepoxy-5,8,5′,8′-tetrahydro-β,β-carotene-3,3′-diol	xe	+	
26	Rhodoxanthin	u − R_3 − u	4′,5′-Didehydro-4,5′-retro-β,β-carotene-3,3′-dione	xk		+
	Rare					
27	γ-Carotene	a − R − g	β,ψ-Carotene	hc	+	+
28	Rubixanthin	a − R − i	β,ψ-Caroten-3-ol	xh		+
29	Gazaniaxanthin	a − R − i	5′-cis-β,ψ-Caroten-3-ol	xh		+
30	3,4-Dehydrolycopene	a − R − b	3,4-Didehydro-ψ,ψ-carotene	xh		+
31	Torulene	b − R − g	3′,4′-Didehydro-β,ψ-carotene	hc		+
32	Neurosporene	a − R_1 − c	7,8-Dihydro-ψ,ψ-carotene	hc		+
33	Torularhodin	f − R − g	3′,4′-Didehydro-β,ψ-caroten-16′-oic acid	xa		+
34	ζ-Carotene	c − R_2 − c	7,8,7′,8′-Tetrahydro-ψ,ψ-carotene	hc		+
35	Dihydroxy-ζ-carotene	e − R_2 − e	1,2,7,8,1′,2′,7′,8′-Octahydro-ψ,ψ-carotene-1,1′-diol	xh		+
36	Dihydroxylicopene	x − R − x	1,2,1′,2′-Tetrahydro-ψ,ψ-carotene-1,1′-diol	xh		+
37	Cryptoflavin	g − R_1 − t	5,8-Epoxy-5,8-dihydro-β,β-caroten-3-ol	xe		

Table 2 (continued)
LIST OF CAROTENOIDS FROM THE LICHEN SPECIES INVESTIGATED

No.	Carotenoid	Structure[a]	Semisystematic name	Group[b]	Occurrence[c] a + c	Occurrence[c] f
38	Flavochrome	h − R_1 − s	5,8-Epoxy-5,8-dihydro-β,ε-carotene	xe		
39	Capsanthin	i − R − v	3,3′-Dihydroxy-β,κ-caroten-6′-one	xk		+
40	Capsorubin	v − R − v	3,3′-Dihydroxy-κ,κ-carotene-6,6′-dione	xk		+
41	Neurosporaxanthin	g − R − w	4′-Apo-β-caroten-4′-oic-acid	ac		+
42	β-Apo-12′-violaxanthal	r − R_4	5,6-Epoxy-3-hydroxy-5,6-dihydro-12′-apo-β-caroten-12′-al	ac		

[a] See Figure 4.
[b] hc, hydrocarbon carotenoids (carotenes); xh, hydroxy xanthophyll; xe, epoxy xanthophyll; xk, ketones xanthophyll; ac, apocarotenoid.
[c] a + c, found in free-living green algae and/or cyanobacteria; f, found in free-living fungi.

The largest group found are the xanthophylls comprising 33 carotenoids; the most common forms are the hydroxyls (12 types), the epoxides (11 types), and the ketones (9 types). Xanthophylls of an acidic character (e.g., torularhodin) are comparatively rare in lichens. The next most common are the carotenes. To date, seven carotenoids of this group have been noted of which β-carotene is the most frequently encountered. The smallest group of carotenoids in lichens consists of apocarotenoids. So far, only two (neurosporaxanthin and β-apo-12′-violaxanthal) have been noted and only in a few species of the families Pertusariaceae, Physiaceae, and Usneaceae. The carotenoids of the other groups have not yet been found in lichens (Table 3.)

With respect to the frequency of occurrence in lichens, carotenoids may be divided into three groups. The carotenoids of the first group, though few in number, are the most commonly encountered. This group includes β-carotene, β-cryptoxanthin, zeaxanthin, lutein, lutein epoxide, astaxanthin, and mutatoxanthin. The second group, which is the largest (19 types), comprises carotenoids frequently found in lichens, usually xanthophylls and two carotenes (lycopene and α-carotene). Finally, the third group consists of 16 carotenoids rarely found in lichens: 4 carotenes (γ- and ζ-carotene, neurosporene, and torulene), 2 carotenoids of the apocarotenal group (neurosporaxanthin and β-apo-12′-violaxanthal), and a number of xanthophylls.

As regards the group of carotenoids found frequently in lichens, all of them have been determined in free-living cyanobacteria and green algae;[3,11-13] most of them have been noted in free-living fungi.[3,14,15] Of the second group (those often found in lichens), most of them have been found in free-living algae (except lycoxanthin, lycophyll, and rhodoxanthin). These latter three and lycopene, echinenone, and canthaxanthin are also encountered in fungi. Of the rarely noted carotenoids, only γ-carotene has been found in free-living cyanobacteria, green algae, and fungi. With the exception of flavochrome, cryptoflavin, and β-apo-12′-violaxanthal, the remaining carotenoids have been found in free-living fungi.

The total carotenoid content of the lichen species studied to date ranges from 0.7 to 94.7 μg/g dry weight.

It should be noted that, in the thalli of some lichens, certain carotenoids may accumulate. This applies in particular to auroxanthin and mutatoxanthin. The latter carotenoid accumulates in large amounts in species of *Caloplaca, Teloschistes,* and *Xanthoria*, giving the thalli of these species a yellow-golden orange color. In such species as *Pseudocyphellaria aurata* and *P. clotharata* in Brazil, auroxanthin constitutes 86.9 to 96.3% of the total carotenoids in the thallus.

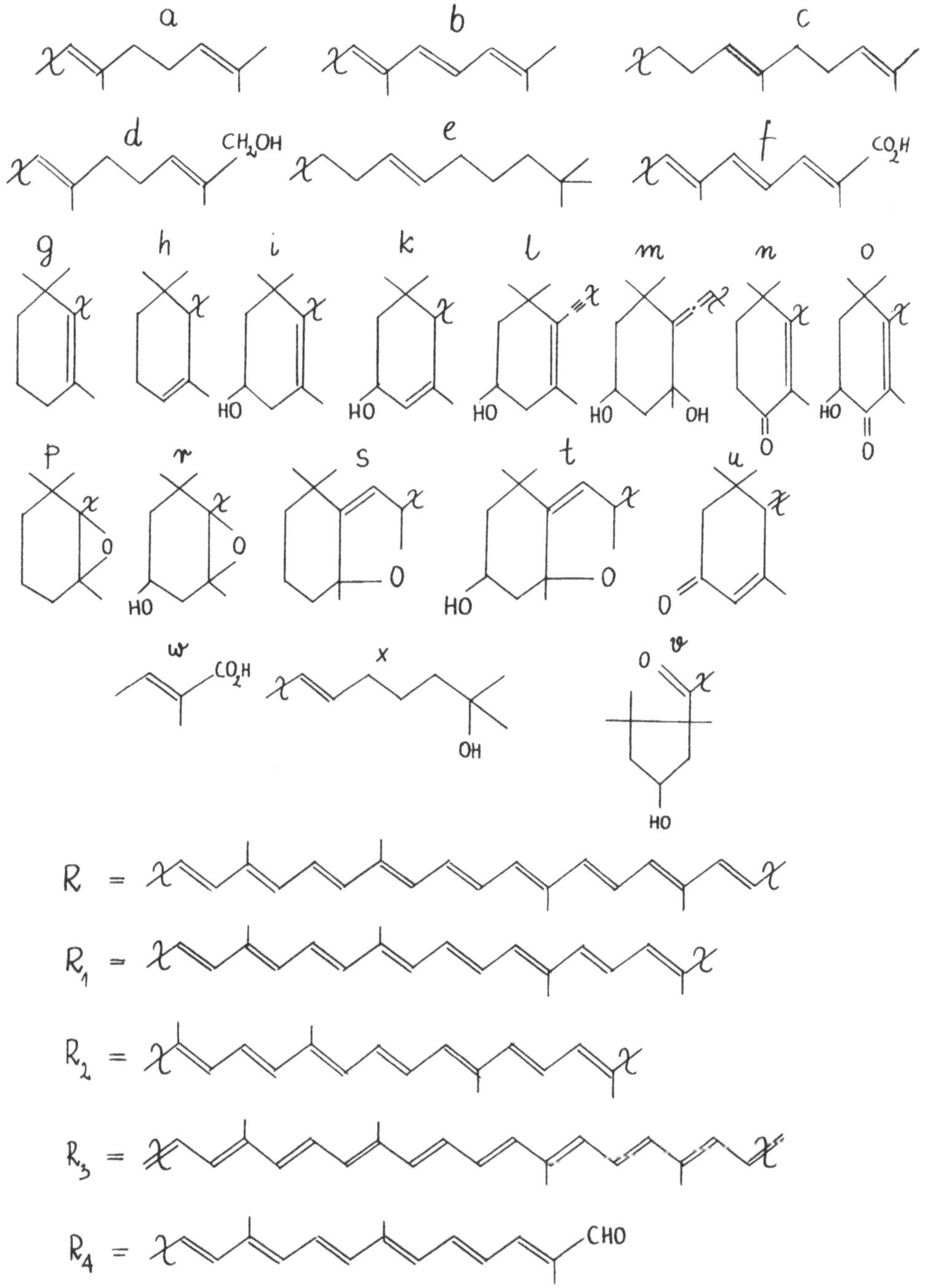

FIGURE 4. Structural features of carotenoids (see Table 2). a to v, end groups; R to R_4, bonds.

The presence of various carotenoids, total carotenoid content, and the predominant carotenoids in the same species of lichen from various geographical regions may vary considerably (Table 4). It may be assumed that ecological conditions have a marked effect on the biosynthesis of carotenoids. Biosynthesis is also dependent on the season of the year. This was demonstrated in studies on lichens in Antarctica where distinct ecological conditions

Table 3
CAROTENOID COMPOSITION IN SOME LICHEN FAMILIES

Family	No. of investigated species	Carotenoids (see Table 2)	Total contents (μg/g dry wt)
Dermatocarpaceae	3	1—3, 5—7, 9, 14—16, 20, 21, 23	0.9—5.9
Stictaceae	1	1—7, 14	15.5
Peltigeraceae	19	1—14, 17, 19—24, 26—28, 32, 34—36, 40	1.3—8.8
Cladoniaceae	43	1—7, 9—18, 20—24, 26, 28—32, 37	1.9—93.3
Stereocaulaceae	10	1—7, 9, 10, 12, 14, 16, 17, 21, 23, 26, 37	1.5—14.8
Umbilicariaceae	3	1—6, 11, 21	1.6—8.0
Pertusariaceae	1	1—3, 7, 23, 24, 41	7.2
Lecanoraceae	1	2—5, 26	8.5
Parmeliaceae	15	1—24, 26—28, 30, 32, 37—40	1.4—40.1
Usneaceae	28	1—24, 26, 28, 30, 32, 33, 41, 42	1.5—31.3
Caloplacaceae	3	1—7, 19, 20, 22, 23, 26, 37	1.5—14.5
Teloschistaceae	11	1—8, 12, 14, 20, 22—26, 29, 30, 37, 38, 40	9.9—94.7
Physciaceae	5	1—6, 12, 14, 17, 18, 20, 21, 23—25, 34, 41	0.7—11.7

were found at different seasons. Studies on *X. elegans* showed that the occurrence of various carotenoids and their relative content varied in different months (Table 5); such findings were also obtained for other lichen species (Table 6). The greatest amount of carotenoids was noted in Antarctic lichen species in spring. It would seem that the presence of the various carotenoids and their total content were affected by the substrate and further by the phase of growth.[17]

The carotenoids have a number of physiological functions; among others, they play an important role in the process of photosynthesis by acting as additional antennae which absorb light rays of different wavelengths than those absorbed by the chlorophylls. They also protect tissues that take part in photosynthesis as well as other tissues from the photooxidizing action of UV light. It is also known that carotenoids participate in phototropism and phototaxis in some lower plants.[1-3] At least two of these roles are applicable to lichens. In all species growing in shaded places, carotenoids play an essential role as antennae which absorb light rays, particularly in the cyanobionts.[16] This can be demonstrated clearly with lichen species such as *Hypogymnia tubulosa* and *Parmelia caperata* (Table 7). Where there is intensive insolation, a marked increase in the carotenoid as well as the chlorophyll content occurs. Furthermore, the carotenoids play a protective role in species found in well-insolated sites.[8] This is especially noticeable in *Xanthoria* (Table 8). When growing in less-insolated places, thalli contained fewer carotenoids than those of the same species in well-insolated sites. In thalli exposed to more sunlight, major increases in the amount of mutatoxanthin were noted.

In the author's opinion, data indicating other physiological functions of lichen carotenoids can be expected in the near future.

Table 4
CAROTENOID CONTENT IN THE SAME SPECIES FROM DIFFERENT LOCALITIES

Species	Locality	Carotenoid (see Table 2)	Major carotenoid (%)	Amount (μg/g dry wt)
Dermatocarpon miniatum	Bulgaria	1, 3, 5, 7, 9, 15, 16, 23	5 (39.9)	5.9
	Israel	3, 5—7, 14, 20	6 (48.7)	4.7
	Greenland	1—3, 6, 21, 23	6 (55.0)	0.9
Cladina alpestris	Poland	2—5, 16, 17, 22	16 (53.3)	11.3
	Lapland	3, 6, 7, 21	7 (62.1)	3.9
C. rangiferina	Poland	1—3, 5, 6, 22	6 (25.1)	9.1
	Greenland	2, 3, 6, 12, 21, 23, 26	3 (53.8)	6.8
	Lapland	2, 3, 6, 7, 14, 23	3 (37.0)	3.9
	Tajmyr (Siberia)	2, 3, 5—7, 9, 20	20 (24.6)	12.2
Cetraria cucullata	Spain	2, 3, 5, 10, 14	3 (46.3)	2.3
	Tajmyr (Siberia)	1, 2, 5—7, 12, 18, 22	5 (30.0)	15.6
C. islandica	Poland	3, 4, 14, 27, 28, 30, 38	28 (30.2)	6.0
	Greenland	1—4, 6, 16, 26	6 (26.6)	4.2
	Lapland	2—4, 13, 17, 18, 20, 23	2 (33.0)	7.5
	Tajmyr (Siberia)	1, 2, 5—7, 10, 12	2 (32.7)	5.4
Xanthoria aureola	Poland	1, 2, 4, 6, 7	7 (94.4)	79.4
	Israel	3—7, 20	6 (29.6)	9.9
X. elegans	Poland	2, 3, 6, 7, 12, 14, 37	7 (42.2)	16.3
	Antarctica	3, 6, 10, 12, 14, 16, 17, 21, 26, 37	7 (32.6)	13.9

Table 5
PRESENCE OF VARIOUS CAROTENOIDS IN *XANTHORIA ELEGANS* FROM ANTARCTICA (PERCENTAGE OF TOTAL)

Carotenoid	October 21, 1980	December 2, 1980	February 3, 1981
β-Carotene	5.4		
α-Cryptoxanthin			5.7
β-Cryptoxanthin	15.4	9.6	5.1
Canthaxanthin	11.6		
Lycoxanthin		15.7	
Lutein epoxide	8.2	23.0	12.7
Zeaxanthin	10.4	18.7	14.7
Adonixanthin			13.2
α-Doradexanthin	13.7		
Astaxanthin		8.1	4.0
Cryptoflavin		3.6	2.4
Mutatoxanthin	32.6	14.8	42.2
Rhodoxanthin	2.7		
Unknown		6.5	
Total contents (μg/g dry wt)	13.9	6.3	3.6

Table 6
TOTAL CAROTENOID CONTENT IN SPECIES FROM ANTARCTICA (μg/g dry wt)

Species	October 21, 1980	December 2, 1980	February 3, 1981
Ramalina terebrata	10.2	8.6	4.8
Usnea antarctica	16.0	8.4	7.4
U. faciata	18.2	15.2	7.3
Caloplaca regalis	15.5	14.5	12.4
Xanthoria elegans	13.9	6.3	3.6
Himanthormia lugubris	18.7	10.9	8.6

Table 7
CAROTENOID CONTENT IN TWO LICHEN SPECIES FROM SHADY AND SUNLIT SITES (μg/g dry wt)

	Site[a]	
Species	Sunlit	Shady
Hypogymnia tubulosa var. *subtilis*	1.49	3.17
H. tubulosa var. *tubulosa*	0.29	1.22
Parmelia caperata	0.82	1.87

[a] For sunlit site, the intensity of sunlight is 4.8 W/m^2; for shady site, the intensity of sunlight is 1.7 W/m^2.

Table 8
CAROTENOIDS AND MUTATOXANTHIN CONTENT IN TWO *XANTHORIA* SPECIES FROM SHADY AND SUNLIT SITES[a]

	Shady site		Sunlit site	
Species	Carotenoid content (μg/g dry wt)	Mutatoxanthin (%)	Carotenoid content (μg/g dry wt)	Mutatoxanthin (%)
Xanthoria fallax	25.1	68.0	71.0	78.6
X. parientina	48.2	58.7	63.4	80.5

[a] For sunlit site, the intensity of sunlight is 6.4 W/m^2; for shady site, the intensity of sunlight is 3.1 W/m^2.

REFERENCES

1. **Isler, O.,** *Carotenoids,* Birkhauser Verlag, Basel, 1971, 932.
2. **Goodwin, T. W.,** *Chemistry and Biochemistry of Plant Pigments,* Academic Press, New York, 1976, 870.
3. **Goodwin, T. W.,** *The Biochemistry of the Carotenoids,* Vol. 1, Chapman & Hall, London, 1980, 377.
4. **Czeczuga, B.,** Investigations on carotenoids in lichens. I. The presence of carotenoids in representatives of certain families, *Nova Hedwigia,* 31, 337, 1979.

5. **Czeczuga, B.,** Investigations on carotenoids in lichens. II. Members of the Usneaceae family, *Nova Hedwigia,* 31, 349, 1979.
6. **Czeczuga, B.,** Investigations on carotenoids in lichens. III. Species of *Peltigera* Willd. *Cryptog. Bryol. Lichenol.,* 1, 189, 1980.
7. **Czeczuga, B.,** Investigations on carotenoids in lichens. IV. Representatives of the Parmeliaceae family, *Nova Hedwigia,* 32, 105, 1980.
8. **Czeczuga, B.,** Mutatoxanthin, the dominant carotenoid in lichens of *Xanthoria* genus, *Biochem. Syst. Ecol.,* 11, 329, 1983.
9. **Czeczuga, B.,** Cartenoids in representatives of hte Cladoniaceae, *Biochem. Syst. Ecol.,* 13, 83, 1985.
10. **Liaaen-Jensen, S.,** Algal carotenoids and chemosystematics, *Mar. Nat. Prod. Chem.,* 1, 239, 1977.
11. **Czeczuga, B.,** Characteristic carotenoids in algae of different systematic position, *Nova Hedwigia,* 31, 326, 1979.
12. **Wettern, M. and Weber, A.,** Some remarks on algal carotenoids and their interconversion into animal carotenoids, in *Marine Algae in Pharmaceutical Science,* Moppe, H. A., Levring, T., and Tanaka, J., Eds., Walter de Gruyter, New York, 1979.
13. **Weber, A. and Wettern, M.,** Some remarks on the usefulness of algal carotenoids as chemotaxonomic markers, in *Pigments in Plants,* Czygan, F., Ed., Gustav Fischer Verlag, New York, 1980.
14. **Valadon, L. R. G.,** Carotenoids as additional taxonomic characters in fungi; a review, *Trans. Br. Mycol. Soc.,* 67, 1, 1976.
15. **Czeczuga, B.,** Investigations on carotenoids in fungi. VI. Representatives of the Helvellaceae and Morchellaceae, *Phyton (Austria),* 19, 225, 1979.
16. **Czeczuga, B.,** The effect of light on the content of photosynthetically active pigments in plants, *Nova Hedwigia,* 35, 371, 1981.
17. **Czeczuga, B.,** unpublished observations.

Chapter IX.C.2

PHYCOBILIPROTEINS

Bazyli Czeczuga

In addition to the photosynthetically active pigments, the chlorophylls and carotenoids, a group of water-soluble pigments, the phycobiliproteins, can be found in thalli of lichens with cyanobacterial symbionts.[1] These pigments also occur in some, but not all, free-living cyanobacteria. Of the algae studied, they have been noted in red and cryptomonad algae.[2]

To date, the presence of 20 types of phycobiliprotein pigments has been established; the greatest variety have been found in the cryptomonads. All phycobiliprotein pigments can be divided into two basic groups. The first group includes phycobiliproteins of the phycoerythrin type (red pigments), and the second group includes those of the phycocyanin type (blue pigments) and allophycocyanins. Of the phycoerythrins, the following types are distinguished: B-phycoerythrin, b-phycoerythrin, C-phycoerythrin, R-phycoerythrin, phycoerythrocyanin (red-blue), and cryptomonad-type phycoerythrins. The second group includes allophycocyanin, allophycocyanin-B, C-phycocyanin, R-phycocyanin, and cryptomonad phycocyanins. Phycobiliprotein pigments differ from one another in absorption maxima and molecular weight, as well as the amino acid composition of the associated proteins.[2] With respect to cyanobacteria, allophycocyanin, allophycocyanin-B, C-phycocyanin, C-phycoerythrin, and phycoerythrocyanin have been found.

Each phycobiliprotein pigment consists of two components — the protein and the prosthetic group which constitutes the chromatophore. In the phycoerythrin group, phycoerythrobilin is the chromatophore; in the phycocyanins, it is phycocyanobilin.[2] These chromatophores belong to linear tetrapyrrols (Figure 1). In the cyanobacterial cells, the phycobiliproteins are localized in special structures called phycobilisomes, which are arranged in an ordered fashion on the surface of the thylakoids. Numerous electron microscopic studies have shown that the phycobilisomes are supramolecular structures that play the role of basic light absorbers.[3,4] Phycobilisomes *in situ* are semidisc-like in shape and each consists of a triangular central part containing three groups of disc-like subunits. From the triangular central portion, three rod-shaped structures branch off radially; the length of these structures is approximately 10 nm. The average number of discs forming the rod-shaped structures varies in the different species of cyanobacteria depending upon the spectral composition of the incident light rays. The triangular part of the phycobilisome contains allophycocyanin, while the rod-shaped structures contain phycocyanin distally and phycoerythrin medially (Figure 2). It should be noted that the migration of the absorbed energy follows a similar direction. phycoerythrin to phycocyanin to allophycocyanin to chlorophyll a.

Phycobiliprotein pigments serve as antennae absorbing rays of a given wavelength from the environment. Whereas chlorophyll a absorbs mainly in the red and blue and carotenoids in the blue-violet, the phycobiliprotein pigments absorb the green (phycoerythrins) and the yellow-red rays (phycocyanins). This contributes to the ability of cyanolichens to inhabit different ecological niches. In addition, they are important in chromatic adaptation, which involves a change in the concentration of the various phycobiliprotein pigments depending on the spectral composition of incident light.

To date, ten lichen species of the Peltigeraceae (eight *Peltigera* spp. and two *Nephroma* spp.) and six *Stereocaulon* spp. (Stereocaulaceae) have been studied; only four phycobiliproteins have been determined (Table 1). Both the total content of phycobiliprotein pigments and the relative percent of each vary depending on the intensity of insolation. In high light intensity habitats, total phycobiliprotein content decreased, while in low intensity habitats,

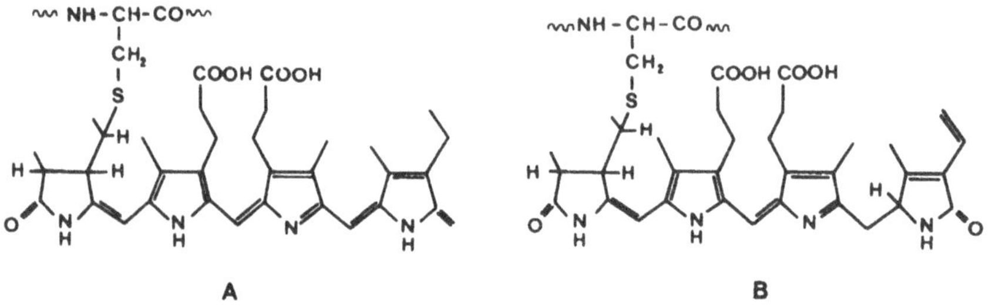

FIGURE 1. Chemical structures of polypeptide-bound chromophores: (A) phycocyanobilin and (B) phycoerythrobilin.

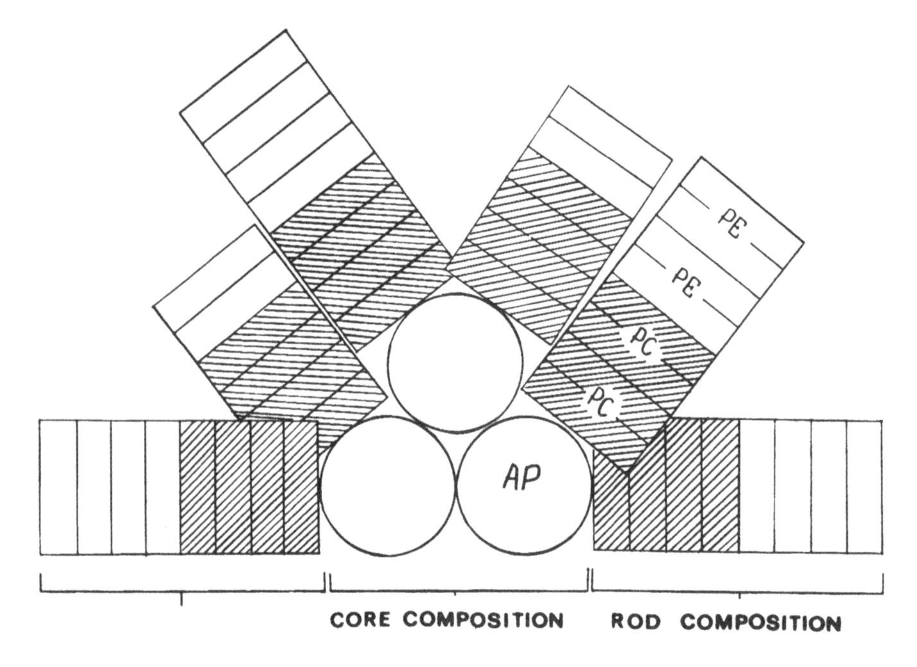

FIGURE 2. Model for a hemidiscoidal phycobilisome. AP, allophycocyanin; PC, phycocyanin; PE, phycoerythrin. (From Gingrich, J. C., Blaha, H. K., and Glazer, A. N., *J. Cell Biol.*, 92, 261, 1982. With modification.)

it increased (Table 2). Studies of the total phycobiliprotein pigment content and the percent content of each pigment in *Peltigera rufescens* thalli in spring, summer, and autumn revealed significant differences (Table 3). In spring and autumn when incident light was less intensive, thalli contained considerably more phycobiliprotein pigments than in summer. Furthermore, in spring and autumn, the percent of C-phycocyanin increased and the C-phycoerythrin decreased; the reverse was observed in summer. It is known that phycocyanin absorbs mainly yellow-red light and phycoerythrins absorb green light. The increase in phycocyanin in *P. rufescens* thalli in spring and autumn and phycoerythrin in summer seems to be related to the differing proportions of these wavelengths of incident light. In early spring and autumn, yellow-red wavelengths predominate, whereas in summer, more green is present. Thus, it would appear that wavelengths of incident light affect the proportion of phycobiliprotein pigments in the thalli.

Table 1
PHYCOBILIPROTEINS CONTENT IN SOME SPECIES

Species	Dates collected	Total content (mg/g dry wt)	Phycobiliproteins (%)		
			AP[a]	PC	PE
Peltigera aphthosa[b]	September 3			40	60
P. canina	March 17	1.536		51.1	48.9
P. degenii	May 16	1.302		42.8	57.2
P. horizontalis	May 20	1.131	41.3	29.4	29.3
P. leucophlebia	October 10	—	—	—	—
P. polydactyla	April 19	1.062		57.7	42.3
P. praetextata	May 16	1.582		34.6	65.4
P. rufescens	May 16	1.650	23.0	41.0	36.0
Stereocaulon alpinum	August 8	0.084	48.0	32.0	20.0
S. arenarium	July 19	—	—	—	—
S. incrustatum	June 21	0.048	44.6	30.4	25.0
S. paschale	August 8	0.075	48.0	32.0	20.0
S. tomentosum	June 28	0.144	46.8	28.1	25.1
S. vesuvianum	August 8	0.027	46.8	31.2	22.0
Nephroma arcticum	August 10	—	—	—	—
N. parile	July 28	0.137	31.6	44.4	24.0

Note: AP, allophycocyanin; PC, C-phycocyanin; PE, C-phycoerythrin.

[a] In *Peltigera* spp., AP was allophycocyanin-B; in *Stereocaulon* spp., it was allophycocyanin.

[b] Cephalodia only.

Table 2
TOTAL PHYCOBILIPROTEINS CONTENT (mg/g DRY WT) IN LICHENS BY DIFFERENT INTENSITIES OF LIGHT

Species	Site[a]	
	Sunlit	Shady
Peltigera rufescens	0.84	1.24
Stereocaulon incrustatum	0.08	0.11

[a] For sunlit site, the intensity of sunlight is 4.[illegible] W/m^2; for shady site, the intensity of sunlight is 2.3 W/m^2.

Table 3
PHYCOBILIPROTEINS CONTENT IN *PELTIGERA RUFESCENS*

Month	Total content (mg/g dry wt)	C-phycocyanin (%)	C-phycoerythrin (%)
March	1.850	52.6	47.4
April	1.754	51.2	48.8
May	1.703	51.5	48.5
June	0.845	43.1	56.9
July	0.640	40.0	60.0
August	0.525	37.7	62.3
September	0.680	42.0	58.0
October	1.540	49.2	50.8
November	2.020	54.0	46.0

REFERENCES

1. **Czeczuga, B.,** Studies of phycobiliproteins in algae. III. Phycobiliproteins in the phycobionts of the *Peltigera* species, *Nova Hedwigia,* 36, 687, 1982.
2. **O'Carra, P. and O'Heocha, C.,** Algal biliproteins and phycobilins, in *Chemistry and Biochemistry of Plant Pigments,* Goodwin, T. W., Ed., Academic Press, New York, 1976, 328.
3. **Gantt, E.,** Phycobilisomes, *Ann. Rev. Plant Physiol.,* 32, 327, 1981.
4. **Gingrich, J. C., Blaha, H. K., and Glazer, A. N.,** Rod substructure in cyanobacterial phycobilisomes: analysis of *Synechocystis* 6701 mutants low in phycoerythrin, *J. Cell Biol.,* 92, 261, 1982.

Section X: Principles of Classification and Main Taxonomic Groups

Chapter X

PRINCIPLES OF CLASSIFICATION AND MAIN TAXONOMIC GROUPS

Josef Hafellner

I. INTRODUCTION

Genera and higher systematic categories were frequently based, in the past, on a single character such as growth form, spore color, or spore septation. Such systematic classifications have been — and many still are — highly heterogeneous. Many older proposals for systematic arrangements of families and genera had to be changed due to an increasing attention to ultrastructure, primarily of asci and ascospores, and to ontogenetic features of fruiting bodies. This process of changes in systematic arrangements, due to more detailed studies by modern means, is still incomplete.

Most of the better-known and accepted families of lichenized fungi have been described in the 19th century, when growth form played an important role in taxonomy of higher categories. Clearly related species and genera, as seen in ascocarp structure and other characters derived from the fruiting bodies, were placed in different families because of different thallus organization (e.g., *Caloplaca — Teloschistes; Rinodina — Physcia; Dirina — Roccella*).

Although in the special case of Teloschistaceae the idea of a homogeneous family comprising all possible growth forms can be found by the 19th century,[1] overemphasizing of thallus organization was not given up generally, probably because such an artificial system was very easy to handle.

Nowadays, attempts of systematic arrangement are based primarily on ultrastructural and ontogenetical characters,[2,3] although they are difficult to analyze without much experience. Practically all suggestions for alterations in systematic arrangements based on differences in ultrastructure have also found support in other respects, e.g., Lecideaceae s. str. with a special ascus type also have specific lichen substances,[4] members of the Porpidiaceae with a certain ascus type and ascospores with a distinct gelatinous sheath are always growing on rocks and a distinction within *Psora*-like genera based on ascocarp characteristics is accompanied by a distinct ecology.[5-7]

II. CRITERIA FOR MODERN CLASSIFICATION

Used today and regarded as important criteria for classification of lichenized fungi are

- Characters derived from asci, mainly the construction of the apices, or basidia, respectively, including the type of dehiscence (see Volume I, Chapter V.A)
- Characters derived from the ascomata or carpophores, respectively, such as ontogeny, types of tissues, carbonization, presence or absence of paraphyses, periphyses, or other kinds of hymenial filaments
- Characters derived from the spores, especially wall construction
- Features of the pycnidia (see Volume I, Chapter V.B)
- Chemistry of lichen substances (see Chapter IX)
- Thallus structure and growth form (the significance of morphological characters is discussed in some detail by Poelt[8]) (see Volume I, Chapter III)
- Ecological features
- Biogeographical characters

The characters discussed here apply only to the lichenized Ascomycotina (Ascomycetes). Essential characters for the classification of higher categories as well as in finding an accurate place in the system for a species in question are derived from the ascus structure. Although known as important for the classification of free-living ascomycetes for several decades, the basic studies in lichen asci had not been carried out before the 1960s by the French school.[9-13] However, some of the details pointed out by the French scientists were not comprehensible to other lichenologists, who declared ascus structure variable and therefore of minor value for taxonomy. After the introduction of electron microscopy to lichenology for taxonomic purposes, ascus structure was proven to be constant at a certain state of ontogenetical development and important for systematics as the type of dehiscence. Many fissitunicate and lecanoralean ascus types have been studied recently by means of light microscopy.[14,15] However, the relatively few detailed transmission electron microscopy (TEM) studies carried out so far[16-26] are not sufficient yet for a complete rearrangement.

Results of ultrastructural studies may also lead to misunderstandings. After the somewhat fissitunicate type of dehiscence was demonstrated in *Rhizocarpon*,[18] a relation with the Lecanidiaceae was argued.[14] Ontogeny of ascomata is essential for distinguishing between several groups (e.g., Peltigeraceae and Lichinaceae),[3,27] all of which can also be circumscribed by morphological-anatomical details. We know too little about many lecanoralean taxa to judge if ascocarp ontogeny will contribute new ideas for the classification of this central group of lichenized fungi. Within the pyrenocarpous lichens, it was demonstrated that both true perithecia and perithecium-like pseudothecia occur.[28] Also, the hysterothecium-like ascocarp is not typical for only one monophyletic group, but can be found in lichenized fungi quite unsimilar in other respects. Another character long used for the classification of families and genera is the formation of different kinds of tissues surrounding the hymenia. We know now that lecanorine and biatorine-lecideine margins may be built in quite closely related genera within one family (e.g., Lecanoraceae) and, on the other hand, a special type of margin may occur in quite unrelated groups.[29] As in nonlichenized fungi, interascal filaments may be of different origin, and different kinds of pigmentation of the tips of paraphyses are a good supporting character for genus delimitation or even the distinction of higher taxonomic levels.[30] Ascospores are also very useful for classification, even if they do not have striking characteristics as in most of the Teloschistaceae,[26] Pyxinaceae (Physciaceae),[31] or Pyrenulaceae. Spore wall composition is important to differentiate between morphologically similar spores of unrelated groups. Septation should not be overemphasized, but more attention should be paid to perispores.

Pycnidia and pycnospores have been used as characters for classification since Nylander[31a] separated *Parmeliopsis* from *Parmelia*. The first systematic study on pycnidia and pycnospores was published at the end of the last century,[32] and has been reexamined recently with modern optical means.[33,34] Several types of pycnidia and pycnospore formation and shape have been distinguished, which may be observed in nearly every possible combination. They give good hints for a natural grouping, but must not be used as the only character because similar types occur in different groups, which can clearly be separated by other characters. Chemistry may be useful in interpreting families and genera (and, of course, taxa of lower taxonomic categories) if the results are seen in connection with morphological features. Pigments especially should not be overemphasized. Almost all discocarp lichens containing anthraquinones have been thought to belong to the same family (Teloschistaceae) until recent times. In reality, pigments of the anthraquinone type have been formed by several not closely related groups (Figure 1). Not all lichen substances are of the same value for taxonomy.

In recent times, growth form is regarded as less important and families with different thallus organization but identical ultrastructural features are normally united. However, it is important to state that ultrastructural features should not always be regarded in the same manner, especially if foliose or fruticose entities cannot be connected with living, more

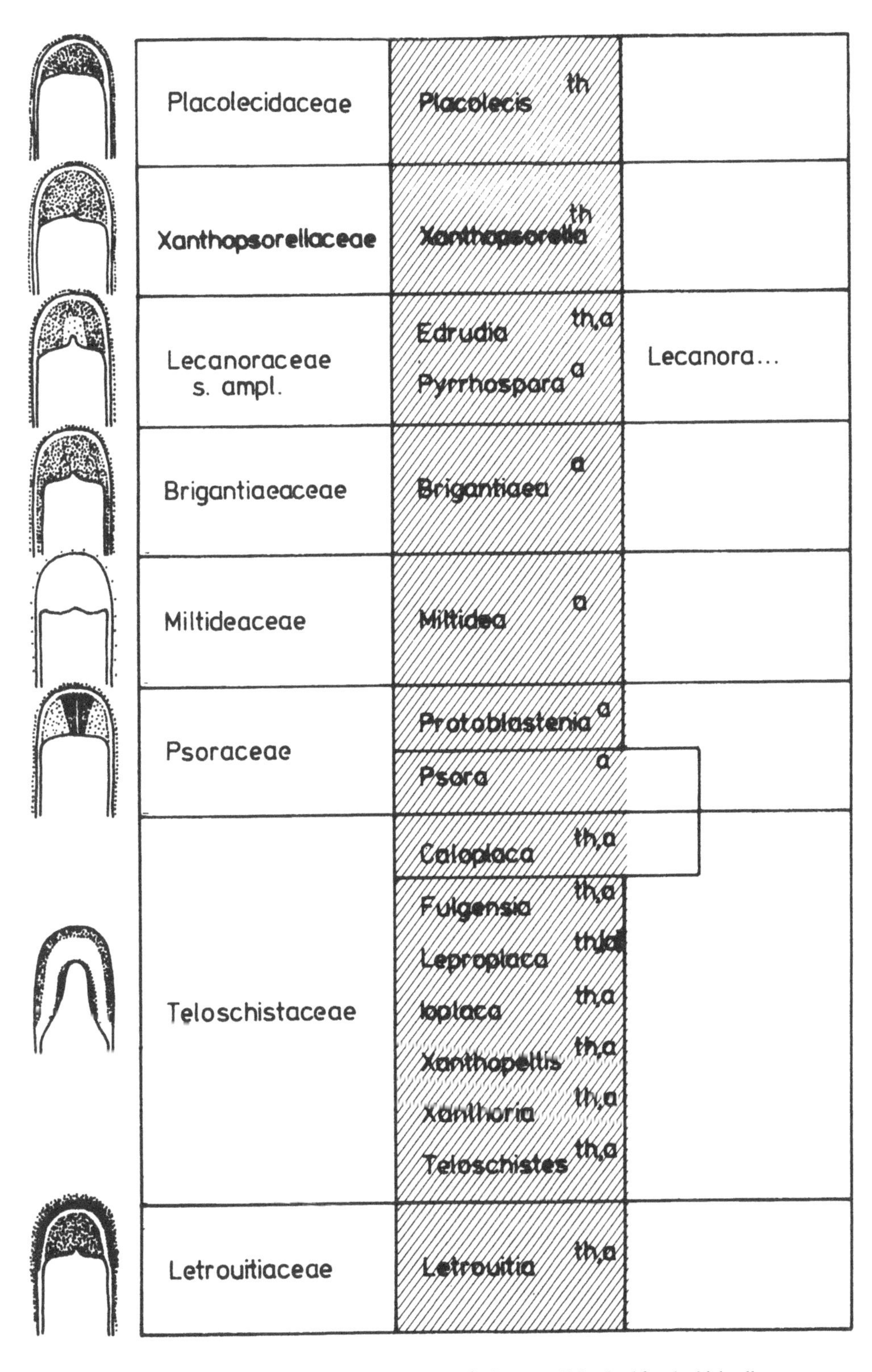

FIGURE 1. The occurrence of anthraquinonous pigments in discocarp lichenized fungi with hyaline ascospores and the corresponding types of asci. th, pigments stored in the thallus; a, pigments stored in the apothecia; amyloid parts of the asci marked by dots (striation indicates the occurrence of anthraquinones in either apothecia or thalli or both).

primitive crustose groups (e.g., Cladoniaceae). For generic circumscription, growth form is a character accepted by nearly all living lichenologists.

Characters derived from ecology and biogeography may be helpful for the separation of more natural groups, especially genera. The Zahlbrucknerian taxonomical concepts are only of historical interest today.

III. RECENTLY INTRODUCED PRINCIPLE OF CLASSIFICATION OF LICHENIZED FUNGI

Since the early 1950s, lichenologists and mycologists have tried to incorporate lichenized fungi in the fungal system,[35] although in many text books the lichens are still kept as a separate entity. This task of incorporation of two systems is very difficult. The basis for modern classification was laid by Nannfeldt,[36] when he separated ascohymenial and ascolocular groups. His ideas were then supported by correlations found between the types of ascocarps and ultrastructural characters in ascus construction.[37]

Although some alterations are necessary [e.g., not only ascolocular, but also some ascohymenial fungi may have fissitunicate asci (see Volume I, Chapter V.A)], Nannfeldt's principles for classification are not questioned. With present knowledge, lichens cannot be maintained as a separate group besides the fungi. One serious consequence was drawn in 1981 by the International Botanical Congress in Sydney. According to the latest issue of the *International Code of Botanical Nomenclature*, names given to lichens refer to the fungal component of the symbiosis. Consequently, the characters derived from the fungus are regarded as most important for the classification of lichens, and both lichenized and nonlichenized fungi have to be classified in one fungal entity depending on the fungal similarities only.

IV. MAJOR TAXONOMIC GROUPS

Lichenized groups are widely distributed in ascomycetes and are much rarer in basidiomycetes. Several of the lichenized groups have no clear relatives in nonlichenized ascomycetes and therefore are classified in separate orders (e.g., Lecanorales and Verrucariales). Other groups normally not so rich in species can easily be attached to well-known orders of nonlichenized fungi (e.g., Dothideales and Tricholomatales). Several proposed suborders (e.g., Peltigerineae, Teloschistineae, and Pertusariineae) have been raised recently to order level, but this probably does not reflect real relationship (Figure 2). Family classification often remains unsatisfactory because many genera have not been studied in detail, so that diagnostic characters regarded as important nowadays are still unknown.

The following orders have been proposed and most of them are generally accepted.[38]

A. Ascomycotina

1. Lecanorales

Comprises the major part of the discocarp lichenized fungi, practically all members lichenized or living on lichens; apothecia are very polymorphous; asci appear generally with thick multilayered walls and many different types of ascus apices; dehiscence is often rostrate, but other types are also known; spores are hyaline or pigmented, unicellular, or septate; all possible growth forms; occupies various substrates in all zones. (See Volume I, Chapter V.A.)

In Lecanorales, many isolated groups are included, most of them with crustose thallus organization (Figures 3 and 4). Some of them are commonly classified as suborders, such as Cladoniineae, Umbilicariineae, and Buelliineae. This is a highly heterogeneous order.

orders with lecanoralean asci; ascus apex normally with tholus; ascomata are apothecia or are derived from apothecia

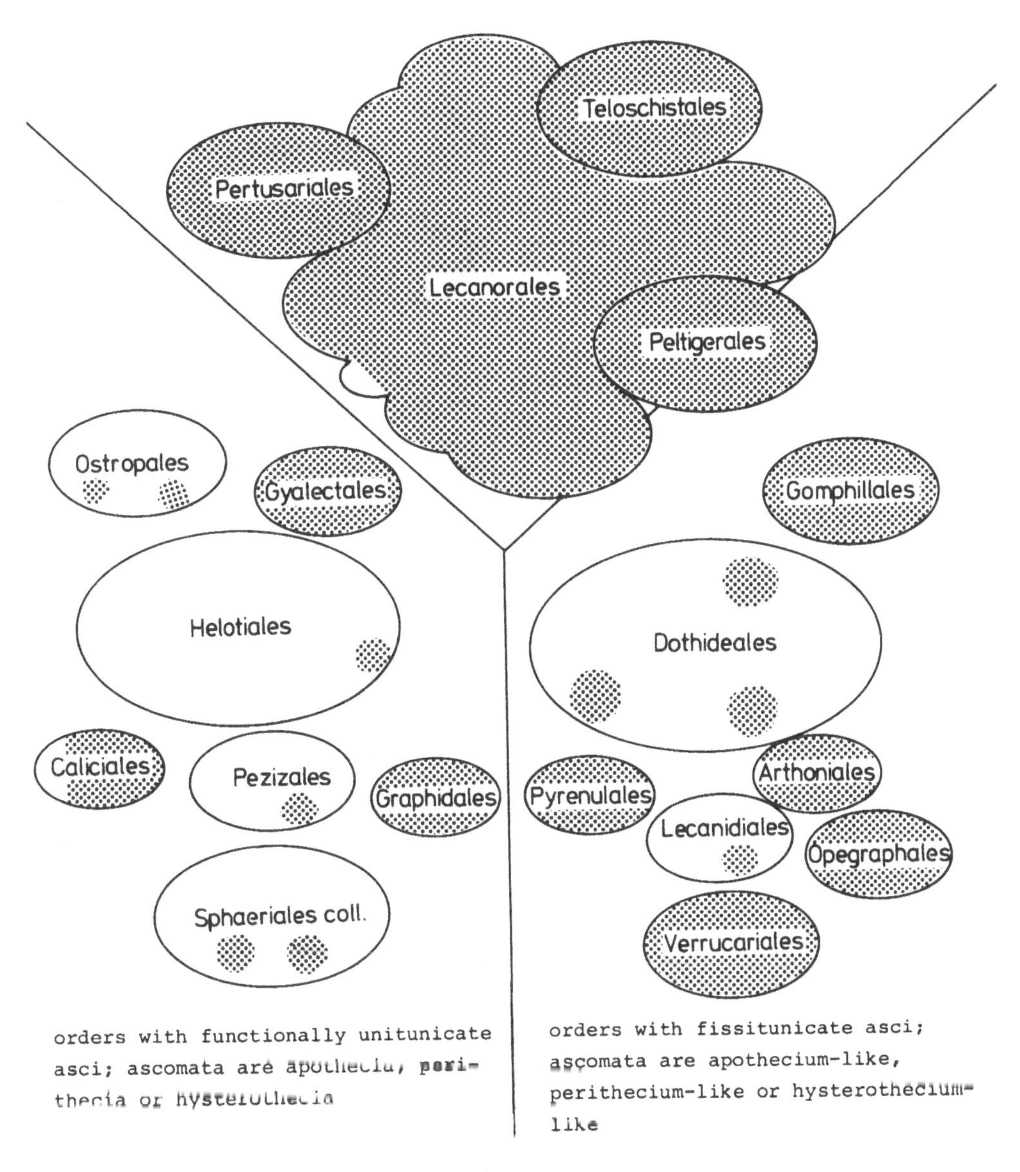

FIGURE 2. The distribution of lichenized fungi in ascomycetes. Orders without lichenized members not included; lichenized groups marked by dots.

Representative families: Acarosporaceae, Alectoriaceae, Anziaceae, Arctomiaceae, Bacidiaceae (including Lecaniaceae and Biatoraceae), Brigantiaeaceae, Candelariaceae, Catillariaceae, Catinariaceae, Cladoniaceae (including Cladiaceae), Coccocarpiaceae, Collemataceae, Crocyniaceae, Ectolechiaceae (including Lasiolomataceae), Eigleraceae, Haematommataceae, Harpidiaceae, Heppiaceae, Heterodeaceae, Hypogymniaceae, Koerberiellaceae, Hymeneliaceae (including Aspiciliaceae), Lecanoraceae, Lecideaceae, Lichinaceae, Lithographaceae, Lopadiaceae, Megalariaceae, Megalosporaceae, Micareaceae (including Helocarpaceae), Miltideaceae, Mycobilimbiaceae, Mycoblastaceae, Orphnio-

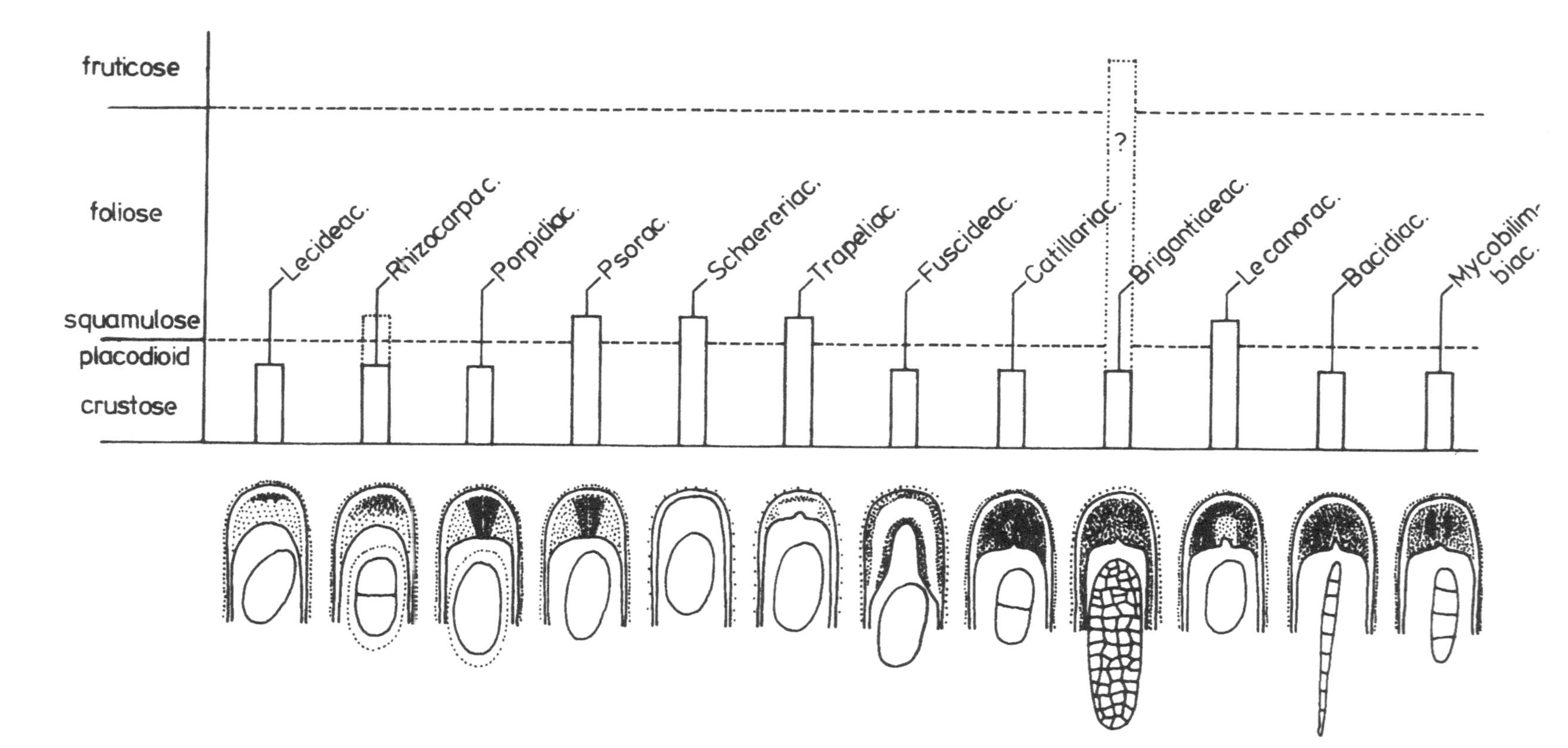

FIGURE 3. Some families of discocarp crustose lichenized fungi with chlorococcoid phycobionts and apothecia surrounded by a proper margin (= the former Lecideaceae coll.) with clearly different types of asci and their growth forms. Amyloid parts of the asci marked by dots.

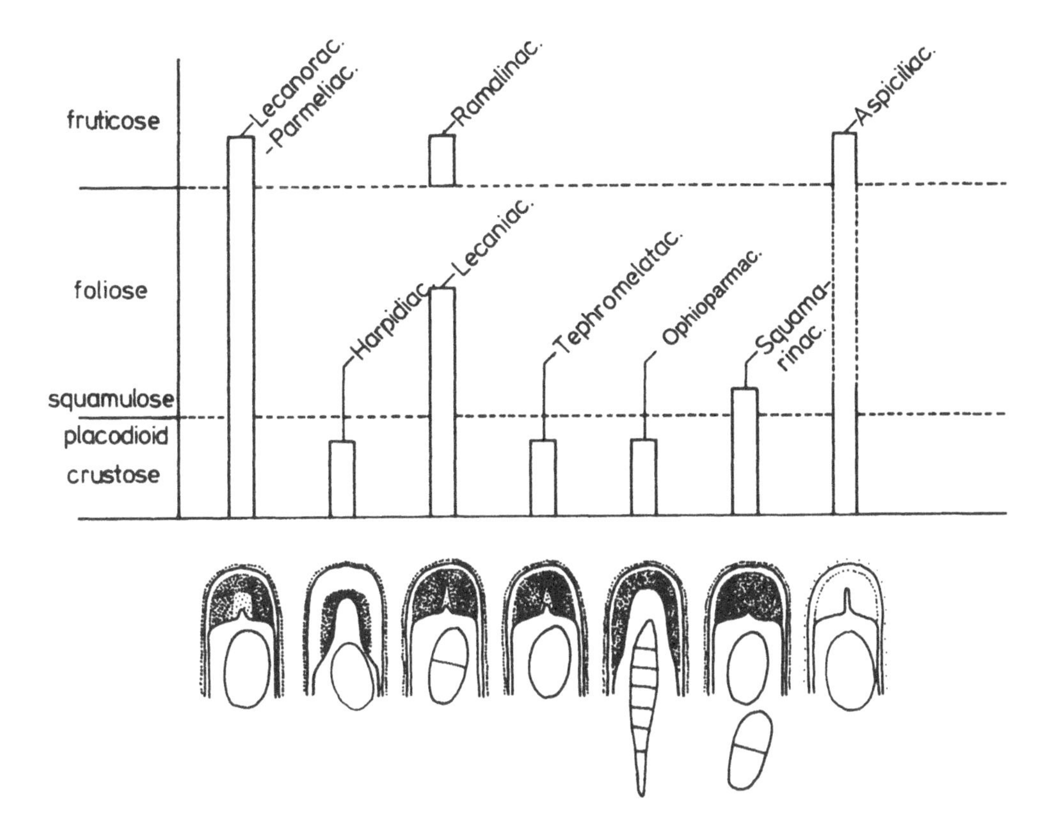

FIGURE 4. Some families of discocarp crustose lichenized fungi with chloroccoid phycobionts and apothecia surrounded by a thalline margin (= the former Lecanoraceae coll.) with clearly different types of asci and their growth forms. Amyloid parts of the asci marked by dots.

sporaceae, Pachyascaceae, Pannariaceae, Parmeliaceae, Pilocarpaceae, Placolecidaceae, Porpidiaceae, Psoraceae, Psorulaceae, Pyxinaceae (including Physciaceae), Ramalinaceae, Rhizocarpaceae, Rimulariaceae, Roccellinastraceae, Ropalosporaceae, Saccomorphaceae (including Trapeliaceae), Sarrameanaceae, Schadoniaceae, Stereocaulaceae, Scoliciosporaceae, Sphaerophoropsidaceae, Squamarinaceae, Tephromelataceae, Thelocarpaceae, Tremoleciaceae, Umbilicariaceae, Vezdaeaceae, and Xanthopsorellaceae.

2. Pertusariales

Apothecia with open discs or perithecia-like and enclosed in thalline warts; asci are thick walled with an amyloid inner wall layer that is thickened at the apex; dehiscence is not fissitunicate; paraphyses anastomosing; spores hyaline, often large and polyenergidic; algae are protococcoid (see Volume I, Chapter II.B), thallus is crustose; occupies arctic to tropical zones on various substrates, but rarely on calcareous rocks.

Representative family: Pertusariaceae.

3. Peltigerales

Apothecia with hemiangiocarpous development; asci are fissitunicate and with an amyloid ring; paraphyses mostly unbranched; spores are septate and often brown when mature; thalli are mostly foliose, often with cyanobionts either in thallus only or in cephalodia (see Volume I, Chapter III); occupies boreal to tropical zones, mostly on humid places on various substrates.

Representative families: Lobariaceae, Nephromataceae, Peltigeraceae, and Solorinaceae; the Lecotheciaceae commonly classified here probably do not belong to this order.

4. Teloschistales

Ascomata are apothecia, asci apically with a thick amyloid external wall layer, tholus present or reduced, dehiscence not fissitunicate, spores often hyaline, polardyblastic or derived from this type, all growth forms, often with anthraquinoide pigments, in all zones on various substrates.

Representative families: Fuscideaceae, Letrouitiaceae, and Teloschistaceae.

5. Gyalectales

Apothecia of hemiangiocarpous origin, with concave discs and a paraplectenchymatic proper margin; asci are thin walled without a tholus, are not amyloid or hymenial jelly faintly amyloid, opening with a porus; spores are colorless, septate, mostly with trentepohlioid algae (see Volume I, Chapter II.B), thallus is crustose; from the temperate to the tropical zone, often in moist habitats.

Representative family: Gyalectaceae.

6. Ostropales

Apothecia with open to punctiform discs, often immersed; wall of the ascocarp is paraplectenchymatous; periphysoids are usually present; asci are not amyloid, thick walled, not fissitunicate; ascospores hyaline, septate, with thin septa; thallus is crustose; many nonlichenized species, epiphytic; occupies mainly tropical but up to the temperate zone.

Representative families with lichenized members: Odontotremataceae and Stictidaceae.[39]

7. Caliciales

Apothecia often with distinct stalk; asci are pseudoprototunicate and hymenial elements forming a mazaedium or asci with thickened apices and spores released through a porus; spores are mostly pigmented and ornamented; thallus is crustose to fruticose; occupies arctic to tropical zones on various substrates; several members are not lichenized.

Representative families: Caliciaceae, Calycidiaceae, Coniocybaceae, Microcaliciaceae, Mycocaliciaceae, Sclerophoraceae, Sphaerophoraceae, and Sphinctrinaceae.[40]

8. Helotiales

Ascocarps are mostly gymnocarpous apothecia, stipitate or sessile; asci are clavate to cylindrical, opening with an apical pore, with or without an amyloid ring; most frequently eight-spored, with true paraphyses; ascospores of various shapes; only a few members are lichenized with crustose to foliose thallus.

Representative family with lichenized members: Baeomycetaceae.[20,41]

9. Pezizales

Ascocarps are usually fleshy gymnocarpous apothecia, rarely leathery to carbonized; asci are cylindrical to clavate, with an apical operculum or rarely opening with an apical slit; true paraphyses of various types; ascospores are always unicellular, hyaline to pigmented; only a few members are lichenized with crustose thallus.

Representative family with lichenized members: Schaereriaceae.[15,41]

10. Graphidales

Ascocarps lirelliform to roundish or resembling perithecia; asci with apically thickened walls, not amyloid, not fissitunicate, opening with an apical pore; paraphysate; with or without periphysoids; spores are septate to muriform with thickened usually amyloid septa; thallus is crustose, occupies temperate to tropical zones.

Representative families: Graphidaceae and Thelotremataceae.

11. Arthoniales

Pseudothecia irregular to roundish; marginal tissues absent or fragmentary; asci are fissitunicate, usually broadly clavate; ascospores are hyaline to pigmented, septate; thallus is crustose, cottony, or living on lichens; occupies boreal to tropical zones; rather closely related to the next order and separation probably not justified.

Representative families: Arthoniaceae and Chrysotrichaceae.

12. Opegraphales

Pseudothecia roundish or lirelliform, solitary or immersed in a stroma, with a distinct excipulum; asci are fissitunicate, spores are hyaline to pigmented, septate; thallus is crustose to fruticose or living on lichens; algae, if present, are mostly trentepohlioid; occupies temperate to tropical zones.

Representative families: Chiodectonaceae, Opegraphaceae, Roccellaceae, and Lecanactidaceae.

13. Gomphillales

Ascomata are apothecia, with a distinct excipulum; asci are fissitunicate and not amyloid; spores are hyaline and usually septate, with hyphophores; occupies mainly tropical zone, but reaches the temperate zone; thallus is crustose.

Representative family: Gomphillaceae (including most of the genera traditionally classified in the Asterothyriaceae).[42]

14. Lecanidiales (Patellariales)

Pseudothecia apothecioid, usually with carbonized excipula; asci are fissitunicate, not amyloid, at least partly with true paraphyses; ascospores are of various shapes; only a few lichenized species with crustose thallus.

Representative family with lichenized members: Arthrorhaphidaceae.

15. Verrucariales

Pseudothecia perithecioid, often carbonized; asci are not amyloid and apically rather thick walled; with uncertain mode of dehiscence; paraphysoids often gelatinizing; periphyses persistent; spores are unicellular to muriform, hyaline to pigmented; thallus is crustose to foliose and with protococcoid algae (see Volume I, Chapter II.B) or living on lichens, without lichen substances; occupies arctic to subtropical zones, usually on rocks or earth.

Representative family: Verrucariaceae.

16. Pyrenulales

Pseudothecia perithecioid and ostiolate, solitary or united in stromata; asci are (arrested) fissitunicate, not amyloid; usually with anastomosing paraphysoids; spores are septate to muriform, often pigmented, with thickened septa; thallus is crustose; occupies mainly tropical but up to the temperate zone.

Representative families: Pyrenulaceae, Trypetheliaceae, and Laureraceae (?).

17. Dothideales

Pseudothecia resembling perithecia; solitary or united in stromata; asci are fissitunicate and usually not amyloid, with paraphysoids or aparaphysate; spores are mostly septate, often thicker on one end; only a few lichenized members with crustose thallus.

Representative families with lichenized members: Arthopyreniaceae, Pyrenotrichaceae, and Mycoporaceae.

18. Sphaeriales coll.

Ascomata are perithecia; asci are not fissitunicate, with different kinds of apices, ring structures are not rare; with or without paraphyses; spores are polymorphous; only a few lichenized members with crustose thallus; a highly heterogeneous order.

Representative families with lichenized members: Trichotheliaceae and Strigulaceae (?).

Several accepted families cannot be assigned to one of the accepted orders, e.g., Moriolaceae, Phlyctidaceae, and Asterothyriaceae s. str. Several even well-known genera have never been found with ascomata and therefore cannot be assigned to a family with certainty, e.g., *Siphula, Thamnolia, Racodium, Cystocoleus*, and *Lepraria*.

B. Basidiomycotina

For classification of these fungi, the system of Jülich[43] has been adopted. For further characters, consult the publications cited.[43,44]

1. *Tricholomatales*

Basidiocarp is annual, terrestrial, lignicolous, or in symbiosis with algae; agaricoid or of other type; with pileus and often with stipe; hymenophore lamellate or of other type; hyphal system is monomitic or sarcodimitic; cystidia often present; basidia hyaline, clavate, with two to four sterigmata; sporée white to cream or ochraceous; spores are hyaline to yellowish, globose to cylindrical, usually smooth; only a few members are lichenized, and thallus of *Botrydina* or *Coriscium* type.

Representative family with lichenized members: Tricholomataceae.

2. *Cantharellales*

Basidiocarp is annual, terrestrial, or in symbiosis with algae, clavarioid or pileate with stipe, membranaceous to soft gelatinous; hymenium is light colored, smooth or folded; hyphal system is usually monomitic; cystidia are mostly absent; basidia are hyaline, narrowly clavate, thin walled, usually with a basal clamp, with one to eight sterigmata; spores are hyaline, of various shapes, mostly thin walled and inamyloid, with apiculus; only a few members are lichenized.

Representative family with lichenized members: Clavariaceae.

3. *Phanerochaetales*

Basidiocarp is annual or perennial, terrestrial, lignicolous, or in symbiosis with algae; resupinate to pileate, sessile to stipitate, membranaceous, ceraceous or woody; hymenial surface is of different types; hyphal system is usually monomitic; cystidia are often present; basidia are elongated, hyaline, flexuous cylindrical or narrowly clavate, without basal clamp, usually with four sterigmata; spores are hyaline, of various shapes, mostly thin walled, smooth and inamyloid; only a few members are lichenized.

Representative family with lichenized members: Dictyonemataceae.

4. *Atheliales*

Basidiocarp is annual, resupinate, effused, pellicular to thin membranaceous; hymenial surface is normally even; hyphal system is monomitic; cystidia are mostly lacking; basidia in clusters, hyaline, clavate, thin walled, usually with four subulate sterigmata; spores are hyaline or greenish-bluish, of various shapes, mostly inamyloid, with apiculus; only a few members living together with algae.

Representative family with lichenized/lichenicolous or algicolous members: Atheliaceae.

REFERENCES

1. **Massalongo, A. B.,** Monografia dei licheni blasteniospori, *Atti I. R. Ist. Sci. Lett. Arti Padova*, 2. ser., 3, app. 3, 5, 1853.
2. **Poelt, J.,** Classification, in *The Lichens*, Ahmadjian, V. and Hale, M. E., Eds., Academic Press, New York, 1974, 599.
3. **Henssen, A. and Jahns, H. M.,** *Lichenes, Eine Einführung in die Flechtenkunde*, Georg Thieme Verlag, Stuttgart, 1973, chap. 13.
4. **Hertel, H.,** Über Saxicole, Lecideoide Flechten der Subantarktis, *Beih. Nova Hedwigia*, 79, 399, 1984.
5. **Schneider, G.,** Die Flechtengattung *Psora* sensu Zahlbruckner — Versuch einer Gliederung, *Bibl. Lich.*, 13, 1, 1979.
6. **Timdal, E.,** The delimitation of *Psora* (Lecideaceae) and related genera, with notes on some species, *Nord. J. Bot.*, 4, 525, 1984.
7. **Wirth, V.,** Phytosoziologie, Ökologie und Systematik bei Flechten, *Ber. Dtsch. Bot. Ges.*, 96, 103, 1983.
8. **Poelt, J.,** Systematic evaluation of morphological characters, in *The Lichens*, Ahmadjian, V. and Hale, M. E., Eds., Academic Press, New York, 1974, 91.
9. **Chadefaud, M., Letrouit-Galinou, M.-A., and Favre, M.-C.,** Sur l'évolution des asques et du type archaeascé chez des Discomycétes de l'ordre les Lécanorales, *C. R. Acad. Sci. Paris*, 257, 4003, 1963.
10. **Chadefaud, M., Letrouit-Galinou, M.-A., and Janex-Favre, M.-C.,** Sur l'origine phylogénétique et l'évolution des Ascomycétes des lichens, *Bull. Soc. Bot. Fr. Mém. Colloq. Lich.*, 1967, 79, 1968.
11. **Chadefaud, M.,** Les asques et la systématique des Ascomycétes, *Bull. Soc. Bot. Fr.*, 89, 127, 1973.
12. **Letrouit-Galinou, M.-A.,** Les asques des lichens et le type archaeascé, *Bryologist*, 76, 30, 1973.
13. **Janex-Favre, M. C.,** Recherches sur l'ontogenie, l'organisation et les asques de quelques Pyrénolichens, *Rev. Bryol. Lichénol.*, 37, 421, 1970.
14. **Eriksson, O.,** The families of bitunicate ascomycetes, *Opera Bot.*, 60, 1, 1981.
15. **Hafellner, J.,** Studien in Richtung einer Natúrlicheren Gliederung der Sammelfamilien Lecanoraceae und Lecideaeceae, *Beih. Nova Hedwigia*, 79, 241, 1984.
16. **Honegger, R.,** *Licht- und Elektronenoptische Untersuchungen an Flechten-Asci von Lecanoratyp*, Dissertation, University of Zúrich, Zurich, Switzerland, 1978.
17. **Honegger, R.,** The ascus apex in lichenized fungi. I. The *Lecanora*, *Peltigera* and *Teloschistes* types, *Lichenologist*, 10, 47, 1978.
18. **Honegger, R.,** The ascus apex in lichenized fungi. II. The *Rhizocarpon* type, *Lichenologist*, 12, 157, 1980.
19. **Honegger, R.,** The ascus apex in lichenized fungi. III. The *Pertusaria* type, *Lichenologist*, 14, 205, 1982.
20. **Honegger, R.,** The ascus apex in lichenized fungi. IV. *Baeomyces* and *Icmadophila* in comparison with *Cladonia* (Lecanorales) and the non-lichenized *Leotia* (Helotiales), *Lichenologist*, 15, 57, 1983.
21. **Bellemère, A. and Letrouit-Galinou, M.-A.,** The lecanoralean ascus: an ultrastructural preliminary study, in *Ascomycete Systematics. The Luttrellian Concept*, Reynolds, D. R., Ed., Springer-Verlag, New York, 1981, 54.
22. **Hafellner, J. and Bellemère, A.,** Elektronenoptische Untersuchungen an Arten der Flechtengattung *Bombyliospora* und die Taxonomischen Konsequenzen, *Nova Hedwigia*, 35, 207, 1982.
23. **Hafellner, J. and Bellemère, A.,** Elektronenoptische Untersuchungen an Arten der Flechtengattung *Brigantiaea*, *Nova Hedwigia*, 35, 237, 1982.
24. **Hafellner, J. and Bellemère, A.,** Elektronenoptische Untersuchungen an Arten der Flechtengattung *Letrouitia* gen. nov., *Nova Hedwigia*, 35, 263, 1982.
25. **Bellemère, A. and Hafellner, J.,** L'appareil apical des asques et la paroi des ascospores du *Catolechia wahlenbergii* (Ach.) Flot. ex Körber et de l'*Epilichen scabrosus* (Ach.) Clem ex Haf. (Lichens, Lécanorales): étude ultrastructurale, *Cryptogamie Bryol. Lichénol.*, 4, 1, 1983.
26. **Bellemère, A. and Letrouit-Galinou, M.-A.,** Le développement des asques et des ascospores chez le *Caloplaca marina* Wedd. et chez quelques Lichens de la famille des Teloschistaceae (*Caloplaca, Fulgensia, Xanthoria*): étude ultrastructurale, *Cryptogamie Bryol. Lichénol.*, 3, 95, 1982.
27. **Henssen, A.,** The lecanoralean centrum, in *Ascomycete Systematics. The Luttrellian Concept*, Reynolds, D. R., Ed., Springer-Verlag, New York, 1981, 138.
28. **Parguey-Leduc, A. and Janex-Favre, M. C.,** The ascocarps of ascohymenial pyrenomycetes, in *Ascomycete Systematics. The Luttrellian Concept*, Reynolds, D. R., Ed., Springer-Verlag, New York, 1981, 102.
29. **Poelt, J. and Wunder, H.,** Über Biatorinische und Lecanorinische Berandung von Flechtenapothecien, Untersucht am Beispiel der *Caloplaca ferruginea* -Gruppe, *Bot. Jahrb.*, 86, 256, 1967.
30. **Kilias, H.,** Revision Gesteinsbewohnender Sippen der Flechtengattung *Catillaria* Massal. in Europa, *Herzogia*, 5, 209, 1981.
31. **Mayrhofer, H.,** Ascosporen und Evolution der Flechtenfamilie Physciaceae, *J. Hattori Bot. Lab.*, 52, 313, 1982.

31a. **Nylander, W.,** Lichenes Scandinaviae, *Not. Sällsk. Fauna Flora Fennica*, 5, 1, 1861.
32. **Glück, H.,** Entwurf zu einer Vergleichenden Morphologie der Flechten-Spermogonien, *Verh. Heidelberg. Naturhist. Med. Ver. N.F.*, 6, 81, 1899.
33. **Vobis, G.,** Bau und Entwicklung der Flechten-Pycnidien und ihrer Conidien, *Bibl. Lich.*, 14, 1, 1980.
34. **Vobis, G. and Hawksworth, D. L.,** Conidial lichen-forming fungi, in *Biology of Conidial Fungi*, Vol. 1, Cole, G. T. and Kendrick, B., Eds., Academic Press, New York, 1981, 245.
35. **Santesson, R.,** Foliicolous lichens. I. A revision of the taxonomy of the obligately foliicolous, lichenized fungi, *Symb. Bot. Upsal.*, 12, 1, 1952.
36. **Nannfeldt, J. A.,** Studien über die Morphologie und Systematik der Nicht-Lichenisierten Inoperculaten Discomyceten, *Nova Acta Reg. Soc. Sci. Upsal.*, 4, 8, 1932.
37. **Luttrell, E. S.,** Taxonomy of the pyrenomycetes, *Univ. Mo. Stud.*, 24, 3, 1951.
38. **Eriksson, O.,** Outline of the Ascomycetes — 1984, *Syst. Ascomyc.*, 3, 1, 1984.
39. **Sherwood, M. A.,** The ostropalean fungi, *Mycotaxon*, 5, 1, 1977.
40. **Tibell, L.,** A reappraisal of the taxonomy of Caliciales, *Beih. Nova Hedwigia*, 79, 597, 1984.
41. **Korf, R. P.,** Discomycetes and Tuberales, in *The Fungi. An Advanced Treatise*, Vol. 4, Ainsworth, G. C., Sparrow, F. K., and Sussman, A. S., Eds., Academic Press, New York, 1973, 249.
42. **Vezda, A.,** personal communication.
43. **Jülich, W.,** Higher taxa of basidiomycetes, *Bibl. Mycol.*, 85, 1, 1981.
44. **Oberwinkler, F.,** Die Gattungen der Basidiolichenen, *Dtsch. Bot. Ges. N.F.*, 4, 139, 1970.

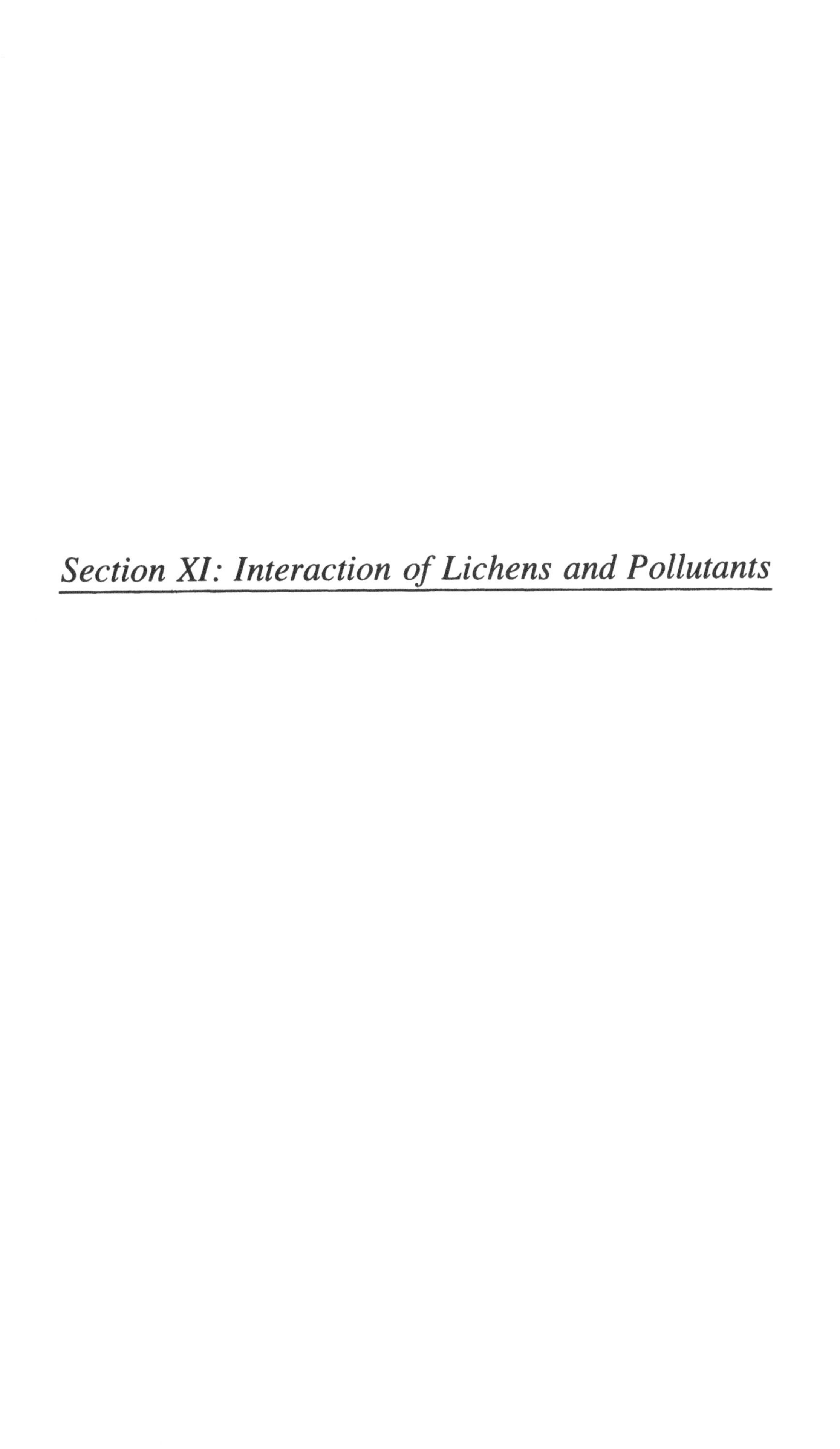

Section XI: Interaction of Lichens and Pollutants

Chapter XI

INTERACTION OF LICHENS AND POLLUTANTS

Margalith Galun and Reuven Ronen

I. INTRODUCTION

The industrial development that started during the second part of the 19th century was accompanied by harmful phenomena such as noise, smoke, and soot. Books by Dickens, Zola, and others refer to this period. With the industrial development, the extension of the mass media, and the rise of the general level of education, people became more aware of environmental problems. This is well demonstrated by the uprise of movements for "return to nature" and by the establishment of government ministries for "environmental problems".

Ecological accidents that reoccur, like the smog "attack" in London which caused the death of 4000 people (1952), the radioactive fallout in the late 1950s from atmospheric nuclear testing, the dioxide leakage at Sevesco, Italy (1976), the "Amos Cadiz" shipwreck on the French shores (1978), the gas accident at Bhopal, India (1984), and the permanent problem of acid rain, have made people more aware of the probability of ecological injuries. This awareness resulted in a demand for more information and for monitoring the environment in which we live.

Thus, monitoring instruments have been established in the vicinity of factories, in urban areas, and at open country sites. However, since such monitoring instruments are very expensive, they can be used at only a limited number of sites and no distribution patterns or comparative data can be achieved.

Concern about these disadvantages led to the search for biological indicators as natural monitors. Among those, lichens were considered most appropriate, already at an early stage. In 1866, Nylander[1] recorded the species of lichens present in the Jardins du Luxembourg in Paris. Thirty years later, in 1896,[2] when he reexamined the species of lichens present in the same Jardins du Luxembourg, he found that all the lichens had disappeared. Nylander asserted that "Les lichens donnent à leur manière la mesure de la salubrité de l'air et constituent une sorte d'hygiomètre tres sensible". Case[3] mentions even earlier studies (from 1859[4] and 1861[5]) on the effect of pollution on lichens. Especially during the last 20 years, numerous studies have been devoted to the assessment of the effect of air pollution on lichens and many reviews have been published.[3,6-12] Since 1974, *The Lichenologist* has devoted, in almost each issue of the journal, a list of publications on air pollution and lichens.

A conclusion that has been drawn from the large volume of research on the interaction between air pollution and lichens is that lichens can be classified into three categories in respect to their relation to air pollution:[12a]

1. Sensitive species, including species on which pollutants have a rather detrimental effect and others with varying degrees of sensitivity. Finally, all become deleted by air pollution.
2. Tolerant species — lichens which are resistant to pollution and remain intact in their native habitat.
3. Species which appear after the major part of the native lichen community has been destroyed by the effect of pollution.

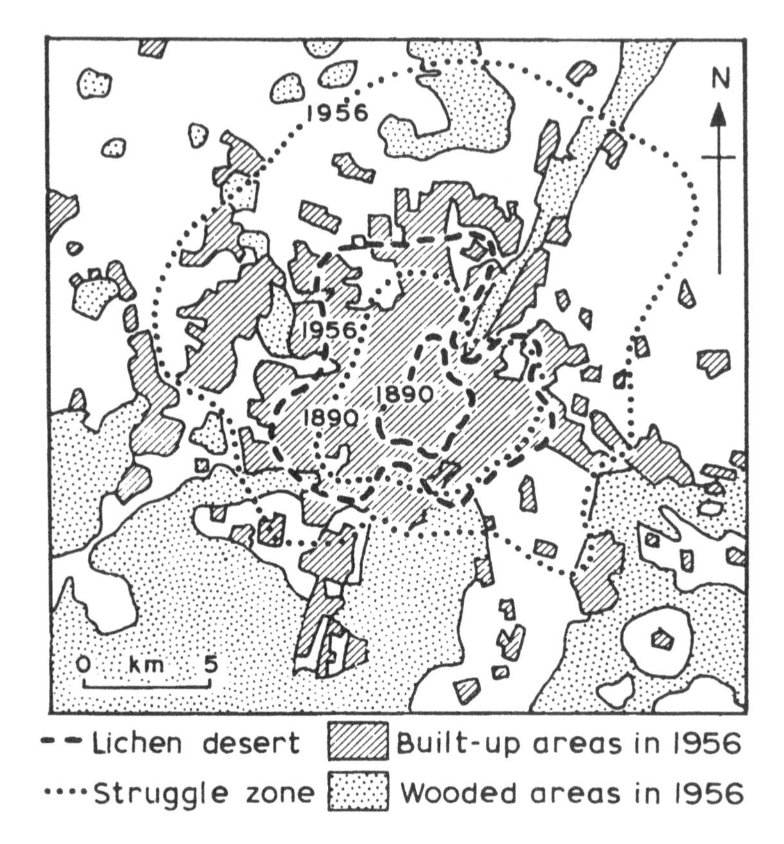

FIGURE 1. Map of Munich showing that damage to epiphytic lichens expanded sixfold between 1890 and 1956. (From Ferry, B. W., Baddeley, M. S., and Hawksworth, D. L., Eds., *Air Pollution and Lichens*, The Athlone Press, London, 1973. With permission.)

II. ECOSYSTEM ALTERATION AND MAPPING

The second important date in the history of lichen biomonitoring is Sernander's[13] research in the Stockholm area. He was the first to introduce the terms "lichen desert" and "struggle zone". He defined the city center, where the tree trunks became bare of lichens, as "lichen desert". Outside the center, a transition zone, where the trunks were only poorly colonized, were defined as the "struggle zone". Beyond that, the zone where lichens thrived was termed the "normal zone". The size of each zone is actually determined by the degree of pollution, the size of the town, and the direction of the wind. A large number of urban zone maps have been established since Sernander's research, e.g., in the Stockholm region,[14,15] in Saarbrücken,[16] in the Copenhagen area,[17] in urban areas in Japan,[18,19] in Zürich,[20] and in Würzburg.[21] When zone maps are established for the same area over an interval of some years, one can observe a change in the distribution of the epiphytic lichens: either an extension of the "desert area" when air pollution is increasing, such as in Zürich[20] or Munich[22] (Figure 1), or a recolonization by lichens, such as in London, when air became cleaner and the sulfur dioxide level fell from 250 to less than 130 $\mu g\ m^{-3}$ after 1960,[23] or in the West Yorkshire conurbation where *Leconora muralis* reinvaded the area after the implementation of the Clean Air Acts (1956 and 1968).[24]

The effect of urban pollution on lichens (the "city effect")[6] has, in fact, been demonstrated at every place where research has been undertaken. The effect does not always cause a complete disappearance of the epiphytic lichens. The ability of some lichens to continue to live in a polluted urban atmosphere has been referred to as "poleotolerance",[25] whereas sensitive lichens are referred to as "poleophobic".

The zonation of urban centers has not been considered by all authors to be due to pollutants. Some authors attributed zonation to changes in climatic conditions such as temperature, and mainly to a decrease in the relative air humidity, a theory known as the "drought hypothesis".[26] After a discussion that lasted 14 years, it has finally been demonstrated[27,28] that air pollution is responsible for the zonation of lichens in urban areas.

The next step in the use of lichens for air pollution assessment was the development of zone maps based not only on the existence of different lichen species, but on frequency and coverage, i.e., quantitative parameters. Trass[29] developed the index of poleotolerance (I.P.), which was formulated as

$$\text{I.P.} = \sum_{1}^{n} \frac{a_1 \times c_1}{C_1}$$

where a_1 = the city tolerance degree of the species; c_1 = the cover ranking value of the species; C_1 = the total cover ranking of all the species studied; and n = the number of species. This value of I.P. ranges between 0 and 10 and enables the establishment of a zonation around towns.

In 1970, LeBlanc and De Sloover[30] recommended the use of another index, the index of atmospheric purity (I.A.P.), which was formulated as

$$\text{I.A.P.} = \frac{1}{10} \times \sum_{1}^{n} Q \times f$$

where n = the number of species found at the station; Q = the ecological index of each species (the number of species found in the vicinity of the species studied at all the stations); and f = the degree of frequency of cover for the species at the particular station. I.A.P. has been used for many areas.[30-33] Deruelle[34] compared these two methods and outlined their advantages and disadvantages. Their common disadvantage is that they introduce a certain factor of subjectivity when the investigator has to determine the degree of city tolerance for I.P. or the degree of species frequencies in the case of I.A.P.

Another approach to biomonitor pollution by lichens is the qualitative one applied by Hawksworth and Rose.[35] They used only epiphytic lichens, distinguished between phorophytes with bark either colonized by algae or not colonized by algae, and chose isolated and well-developed trees. Their mapping is based on the presence or absence of lichen species. These species are categorized in ten different groups. For each kind of phorophytic bark, the mean value of SO_2 concentration in the air is assigned for each group. For example, *L. conizaeoides* is found all along the trunks with bark colonized by algae when the SO_2 concentration is about 125 μg m^{-3} (0.04 ppm). *Usnea ceratina* will be found on the same trees only when the concentration of SO_2 is about 35 μg m^{-3} and *U. florida* will be found if the concentration is less than 30 μg m^{-3}. Studies based on the method of Hawksworth and Rose were undertaken by Hawksworth and Rose[35] for the whole of England and by Deruelle[36] in the Mantes area (France).

Gilbert[37] compared the distribution of lichens around the town of Newcastle-upon-Tyne with the level of SO_2 as measured by chemical monitors, and established a biological scale for estimation of air pollution. This scale includes six different groups of vegetation, each one corresponding to a certain level of SO_2 concentration. Johnsen and Söchting[17] compared the distribution of lichens and the winter average of air pollution measured by 23 monitoring stations in the Copenhagen area and distinguished 5 different classes corresponding to 5 degrees of air pollution. Trass[38] correlated the I.P. in the Estonian SSR to the concentration of SO_2 in the atmosphere. For example, 2 < I.P. < 5 corresponds to a SO_2 concentration of 10 to 30 $\mu g\ m^{-3}$ and I.P. > 9 indicates that the concentration of SO_2 varies between 100 and 300 $\mu g\ m^{-3}$. According to Trass, *L. conizaeoides* appears when the SO_2 concentration is higher than 300 $\mu g\ m^{-3}$.

Although the environment in urban and industrial areas usually contains a mixture of airborne pollutants, which may have synergistic or antagonistic effects, the main effect on lichens has been attributed to SO_2. In fact, in some research reports, "atmospheric pollution" is synonymous with "SO_2 pollution".

III. TRANSPLANTATIONS

Another way to use lichens as biomonitors is to transplant them from a clean place, generally a rural one, to polluted areas, where the natural lichen vegetation has totally disappeared. Brodo[39] transplanted epiphytic lichens with a part of their substrate. Disks of bark were sawed off with a circular saw, transferred to holes of the same diameter, and prepared in trunks of trees of the same species in the polluted area. After a certain exposure time, the lichens were examined. A slight variation in Brodo's method was introduced by Schönbeck,[40] who fixed the disks on a board with special cement and could thus transplant lichens to places lacking trees. Many studies were carried out with lichens on transplanted twigs.

If transplanted to a rural zone at different distances from a determined source of pollution, such as copper smelters,[41,42] power plants,[43-45] an oil refinery,[46] a zinc factory,[47] or steel factories,[48] the effect of specific pollutants on the transplants may indicate the expansion and relative level of it.

Mapping, based on lichen surveys, and criteria such as frequency, coverage, diversity, and abundance, need professional knowledge in lichenology and are influenced by subjective assessments. In order to complement or replace these methods, more objective parameters are continuously attempted. These are usually carried out by analyzing a single lichen species, either native or transplanted, as described in the following section.

IV. MORPHOLOGICAL AND CYTOLOGICAL CHANGES

Pollution-induced changes, which have been suggested as monitoring phenomena, are those that become apparent during the process of deterioration and eventual deletion of the sensitive species. The information obtained enables various levels of pollution to be asserted.

Hypogymnia physodes (= *Parmelia physodes*) has a smooth, grey thallus with large lobes when growing in clean areas, whereas surface cracks, brown or black thallus discolorations, and small lobes appear in thalli exposed to pollution.[10] Electron micrographs first revealed degeneration of cell organelles in the algal cells of *H. physodes* transplants exposed to SO_2, NO_x, and fluorides. After longer exposures, vacuolization and the appearance of vacuolar aggregates were seen in the fungal cells.[49]

The injury of *Parmelia sulcata* transplanted on disks[39] and exposed to fluoride released from an aluminium factory could be graded on the basis of thallus discoloration and detachment from the substrate, changes in soredial and rhizines structure, and plasmolysis of the algal cells in relation to periods of exposure. The study was carried out by periodic photographic assessments and microscopic observations.[50] Gilbert[37,51] observed a massive decrease in size of the fruticose lichens *Evernia prunastri* and *Usnea subfloridiana* in response to pollution along a transect from the outskirts towards the center of Newcastle-upon-Tyne.

The algal cells of a lichen are the primary target of pollution-induced damage. Pigment changes that occur in the algal cells have been revealed by fluorescence microscopy and quantified by microfluorometry.[52] Healthy cells emit a powerful red color; damage results in a gradual change through brown, orange-yellow, and finally white fluorescence. Le Blanc and Rao[53] correlated injury symptoms of *P. sulcata* and *P. milligrana* transplants quantitatively to the level of SO_2 prevailing at five exposure sites in the Sudbury region. The symptoms observed during a six months period were progressive: (1) Color changes from greyish-green to whitish-brown; (2) appearance of a waxy, water-insoluble material cover of the thallus; (3) partial detachment of the thallus from its substrate; and (4) plasmolysis and death of the algal cells.

Accumulation of starch around the pyrenoid was interpreted as an effect of SO_2 on the *Trebouxia* cells of *Parmelia caperata.*[54]

The injury symptoms are more conspicuous in foliose and fruticose species, since most of the crustose lichens are more resistant to air pollution stress.[55]

V. METABOLIC AND PHYSIOLOGICAL PROCESSES AFFECTED BY POLLUTION

The processes examined experimentally and on field material were photosynthesis, respiration, protein and lipid biosynthesis, N_2 fixation (in cyanolichens), pigment breakdown, ATP levels, and potassium efflux.

The techniques employed for photosynthesis and/or respiration measurements were by means of CO_2 gas exchange with an IR gas analyzer (IRGA); ^{14}C incorporation (using $^{14}CO_2$ or $NaH^{14}CO_3$); oxygen evolution measurements, and, recently, photoacoustic spectrometry (PAS). All lichens collected in the field and used for laboratory experiments first have to be revived (or preconditioned). However, the revival duration in the different investigations ranged from less than 1 hr to several days, and were also carried out under different light, temperature, and moisture regimes. It is essential to determine the optimal conditions for each species prior to experimentation.

Türk et al.[56] tested the effect of exposure to 0.5, 1.0, 2.0, and 4.0 mg SO_2 per cubic meter air on 12 different lichen species and detected rather striking species-specific differences in net photosynthesis and dark respiration. All lichen samples were revived for 6 to 8 days before the exposure experiments were carried out, then exposed for 14 hr to the respective concentrations and the effects measured at 7-day intervals for 3 to 4 weeks. No effect could be discerned in *Xanthoria parietina* when exposed to 0.5, 1.0, and 2.0 mg SO_2 per cubic meter air. A slight reduction of net photosynthesis occurred immediately after exposure to 4.0 mg/m^3 SO_2, but this soon recovered to the initial level; 4.0 mg/m^3 raised slightly dark fixation, but 2.0 mg/m^3 had no effect on respiration. The lower concentrations also had no effect on *Lasallia pustulata*. However, 4 mg SO_2 per cubic meter air caused an irreversible 25% reduction in CO_2 fixation. Exposure to 1 and 2 mg/m^3 SO_2 caused a 187 and 214% raise of dark respiration, respectively. The higher dosage had almost no effect. The immediate reaction of CO_2 uptake to 0.5 and 1 mg/m^3 SO_2 exposure by *P. saxatilis* was a reduction of approximately 50%, which gradually recovered and reached, at the end of the experiment, higher net photosynthesis values than the initial ones. (An enhancement of net photosynthesis when exposed to low concentrations of SO_2 was also observed with other species and other techniques.[57-59]) Both 2 and 4 mg/m^3 SO_2 caused irreversible damage, expressed by a 20 and 50% CO_2 uptake reduction, respectively; 0.5 to 2 mg had no effect on dark respiration, and treatment with 4 mg had a permanent effect on it. Carbon dioxide uptake values in *Lobaria pulmonaria* dropped to 60% of the initial value when treated with 1 mg/m^3 SO_2; to 40% when treated with 2 mg; and with 4 mg, the compensation point was reached. After a gradual recovery, photosynthesis reached 85% (1 mg), 75% (2 mg), and 55% (4 mg) of the normal values. The photosynthesis intensity of *L. pulmonaria* was irreversibly damaged, even by the lower values of SO_2, whereas *X. parietina* survived all treatments including the exposure to 4 mg/m^3 SO_2 under the same experimental conditions and showed no permanent impairment on CO_2 gas exchange.

The authors[56] arranged the investigated lichens in the following decreasing order of resistance: *X. parietina, P. scortea, P. acetabulum, H. physodes, P. saxatilis, Platismatia glauca, L. pulmonaria, Parmelia stenophylla, and E. prunastri.*

Damage to net photosynthesis is usually in correlation with the effect on respiration. Severe damage of photosynthesis is generally accompanied by a marked reduction in respiration. However, a noticeable increase of respiration is often the immediate, temporary reaction to SO_2 exposure, similar to the reaction of lichens to other stress impositions.[60,61]

In all investigated cases, the sensitivity of the lichens to either pollutants or temperature stress was closely related to their moisture status. Dry thalli may survive higher concentrations of contaminants, as well as extreme temperatures (see Volume 2, Chapter VII.B.2), without being damaged. However, hydration rate for optimal activity is species dependent and varies from one species to the other and sometimes even in samples of the same species from different environments. Recovery also does not take place in dry thalli.

The deleterious effect of SO_2 is also pH dependent. At lower pH values (below 4), it is much more toxic than at higher values (above 5). The pH of the medium in experiments with aqueous sulfur dioxide thus has to be properly adjusted. The toxicity is not a pH effect, since solutions acidified with HCl are less toxic than solutions of the same pH supplemented with SO_2.[56]

A series of experiments[106] in which the effect of bisulfite on the ATP content, chlorophyll/phaeophytin ratio, and K^+ efflux was examined in *X. parietina* and *Ramalina duriaei* revealed further evidence of the high resistance of *X. parietina*. Both lichens were exposed to various concentrations of bisulfite for different periods of time, and the effects were compared. K^+ efflux was significantly higher in *R. duriaei* than in *X. parietina*, and was initiated at a lower concentration of bisulfite. ATP content of both lichens decreased, but the decrease of ATP content in *R. duriaei* was caused by much lower concentrations of bisulfite. Similarly, the extent of chlorophyll degradation was much lower in *X. parietina*. The same parameters were examined in *X. parietina* from polluted and unpolluted sites and almost no differences could be detected between samples from the polluted environment and clean sites.

The experiments performed by Malhotra and Khan[62] with the sensitive species *E. mesomorpha* were also carried out with relatively low concentrations of gaseous SO_2. Their main objective was to examine the effect of SO_2 on the biosynthesis of the proteins and lipids by measuring incorporation of [U-^{14}C] leucine and [1-^{14}C] acetate, respectively. They also measured the effect of SO_2 on photosynthesic CO_2 fixation by incubating the lichen tissue in $NaH^{14}CO_3$. Their experiments showed that increasing the length of exposure of *E. mesomorpha* to 0.1 ppm SO_2 resulted in progressive reduction of protein biosynthesis and photosynthetic CO_2 fixation. Protein synthesis appeared to be more sensitive to SO_2 than did CO_2 fixation, especially after longer exposures. Two days of exposure to 0.1 ppm SO_2 resulted in about 15% inhibition in the rates of both processes. The inhibitory effect of 0.34 ppm SO_2 was much faster and more pronounced, and protein biosynthesis was inhibited more severely than CO_2 fixation. The extent of recovery of the two metabolic processes differed when the fumigated lichens were transferred to a SO_2-free atmosphere, but appeared only after exposure to the lower SO_2 concentrations. Lipid biosynthesis was similarly affected and partial recovery became apparent even after 50% inhibition of lipid biosynthesis caused by 3 days exposure to 0.34 ppm SO_2. The authors assumed that lichens may also recover under field conditions when SO_2-free periods follow exposures to low levels of SO_2 of a limited duration.

Incubation of *Trebouxia* cells in sodium sulfite solutions affected proline dehydrogenase and caused an increase in alanine and a decrease in proline.[63]

PAS was introduced recently for the in vivo assessment of photosynthetic parameters in lichens[64,65] and applied to air pollution studies.[58,59,66] The photosynthetic responses in *R. duriaei* treated with bisulfite and the long-term effect of total air pollution in the environment

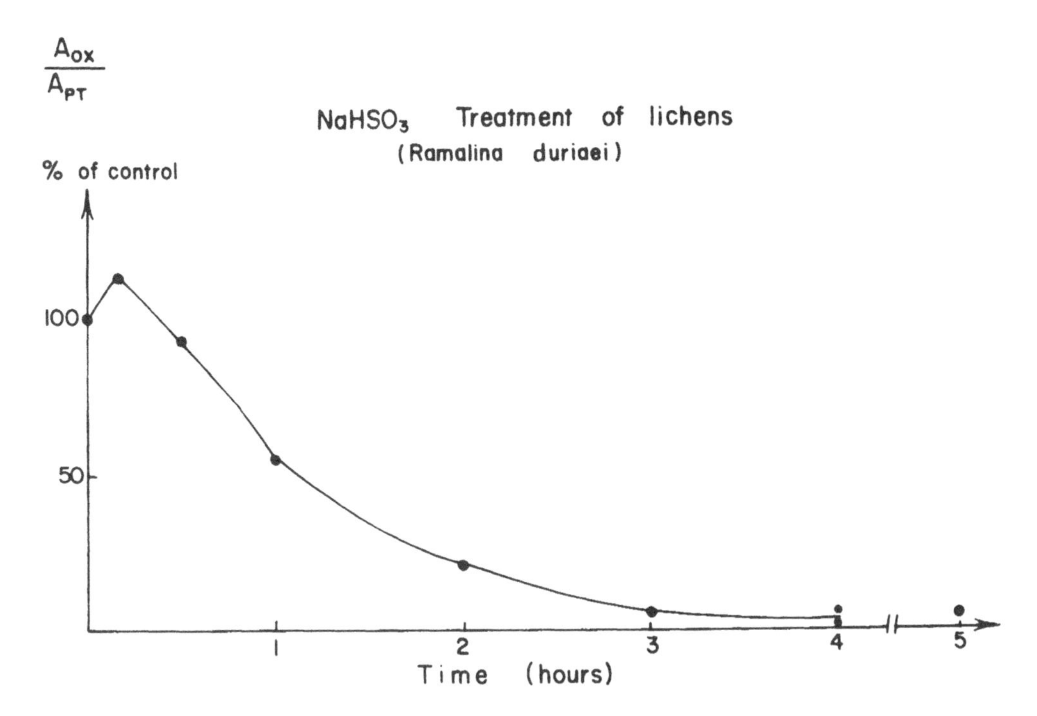

FIGURE 2. Oxygen evolution yield A_{ox}/A_{pt} as a function of incubation time in 5 mM $NaHSO_3$ in the presence of light (8 W/m^{+2}, 400 to 700 nm). A_{ox}, oxygen evolution; A_{pt}, photothermal signal.

on the photosynthetic apparatus of *R. duriaei* transplanted to rural, suburban, and urban sites were measured by this technique.

The kinetics of photosynthesis inhibition by 5 m*M* bisulfite (Figure 2)[67] showed that inhibition is not immediate; incubation for a short period resulted in a temporary increase of the oxygen yield above the control level. This phenomenon was also observed in other lichens.[57,67,68] The quantum yield of oxygen evolution was reduced by about 40% after a 1-hr bisulfite treatment and the quantum yield of energy storage decreased by about 75%. After 4 hr, oxygen evolution and energy storage were totally inhibited. The oxygen evolution quantum yield, reflecting most rate-limiting reactions in the noncycling electron flow, was damaged to a lesser extent than the quantum yield for the energy storage. It seems, therefore, that other reactions (besides the linear electron transport chain) which store photochemical energy were strongly affected by the bisulfite treatment.

The quantum yield of oxygen evolution (A_{ox}/A_{pt}) in transplanted lichens, measured photoacoustically, was in correlation with chlorophyll degradation (Table 1), and the latter was in agreement with the levels of air contaminants at the monitoring sites (Table 2).[66,84]

Chlorophyll content and chlorophyll degradation are also parameters used for measuring the effect of pollutants on lichens. Almost all the methods employed for chlorophyll extraction and estimation of higher plants have been adopted for lichens. However, many of the lichen substances may have an acidifying effect during the extraction and analysis process[69] and may yield incorrect chlorophyll/phaeophytin proportions, and some of these substances also overlap with chlorophyll in their absorption spectra.

Some of the lichen substances can be separated by Sephadex columns.[107] Several short washings with absolute acetone prior to extraction and the addition of carbonate during the extraction reduced the interference of phaeophytinization.[69] Thus, species containing large amounts of lichen substances, particularly those with free carboxyl groups, should be avoided for chlorophyll extraction.

A complete and rapid extraction of the photosynthetic pigments of *R. duriaei* thalli was

Table 1
THE OD 435/OD 415 AND A_{ox}/A_{pt} VALUES OF *R. DURIAEI* SAMPLES SUSPENDED AT THE MONITORING STATIONS (DECEMBER 1981 TO DECEMBER 1982[66]

Location	OD 435/OD 415 $\bar{x} \pm SD$ (10 samples)	A_{ox}/A_{pt} $\bar{x} \pm SD$ (the same 10 samples)
HaZorea	1.43 ± 0.05	1.71 ± 0.68
Nahsholim	1.40 ± 0.03	1.39 ± 0.92
Hedera	1.34 ± 0.11	1.23 ± 0.64
Pardes Hana	1.32 ± 0.10	1.43 ± 0.60
Tel Aviv University (old campus)	1.32 ± 0.20	1.23 ± 0.61
Ofer	1.28 ± 0.07	1.47 ± 0.71
Shemurat Allon	1.24 ± 0.15	0.74 ± 0.51
En HaShofet	1.23 ± 0.18	0.98 ± 0.54
Givat Ada	1.19 ± 0.20	0.90 ± 0.70
Ma'agan Mikhael	1.12 ± 0.25	0.65 ± 0.68
Bet Eliezer	1.07 ± 0.31	0.75 ± 0.69
Kefar HaYarok Junction	0.92 ± 0.24	0.35 ± 0.58

Note: $r = 0.895$; $p < 0.01$.

achieved with dimethyl sulfoxide (DMSO). The procedure is simple; small samples are sufficient for reliable assays; and no grinding or homogenizing is necessary. The ratio of the optical density of the DMSO pigment extract at wavelengths of 435 and 415 nm was used to evaluate chlorophyll degradation in *R. duriaei* after treatment with bisulfite and in transplanted samples.[70] Table 1 shows the correlation between chlorophyll degradation and the quantum yield of oxygen evolution in transplanted *R. duriaei* samples. However, oxygen quantum yield appears to be a more sensitive parameter of pollution damage.[59]

Chlorophyll degradation is one of the later stages in the damage of the photosynthesis apparatus. Even irreversible damage to photosynthesis is not necessarily correlated with destruction of chlorophyll.[55] For example, photosynthetic ^{14}C fixation by *Umbilicaria mühlenbergii* was reduced by almost 90% within 15 min of exposure to aqueous SO_2 (75 ppm, pH 3), whereas no change occurred in the chlorophyll pigments.[71]

Relatively little is known on the effect of SO_2 on N_2 fixation in cyanolichens. Hällgren and Huss[72] showed that N_2 fixation in *Stereocaulon paschale* was more susceptible to bisulfite treatment than photosynthesis in this lichen. N_2 fixation was also more sensitive to pH changes than photosynthesis. An inhibition effect by SO_2 on the rate of N_2 fixation was also reported for *Collema tenax* and *Peltigera canina*,[73] both containing *Nostoc* as their cyanobiont.

Another effect encountered by SO_2 exposure is membrane damage, which becomes apparent by potassium ion leakage from the thallus. This loss is in correlation with the duration of the exposure period and occurs at early stages of damage (Table 3).[71] After a 24-hr recovery period, K^+ efflux was near normal for samples exposed for 15 min to SO_2, but remained high for those exposed for 60 min.

Tomassini et al.[74] exposed *Cladonia rangiferina* to various concentrations of aqueous SO_2 for different periods of time and measured the effect on K^+ release and (in separate experiments) ^{14}C fixation rates. The threshold values (in which no K^+ efflux occurred above the control levels) calculated from these experiments for K^+ efflux were in good agreement with the threshold values evaluated from the ^{14}C fixation rates. These authors also calculated extrapolations for threshold values in polluted areas.

Potassium ions and other electrolites change the conductivity of deionized, distilled water. Changes in conductivity of aqueous solutions in which lichens were immersed were therefore suggested as a sensitive and practical biomonitoring system,[75] and have been used to assess the degree of sensitivity of a number of lichen species.[76]

Table 2
ELEMENT CONTENT AND OD 435/OD 415 VALUES (MEANS AND STANDARD DEVIATIONS OF *R. DURIAEI* SAMPLES SUSPENDED AT THE MONITORING STATIONS, JULY 1980 TO JULY 1981)

												OD 435/OD 415	
Location	Br	Sr	Pb	S	Zn	K	Fe	Ca	Ti	Cl	P	No. of samples	Mean
HaZorea	5 ± 0	14 ± 1	13 ± 4	2312 ± 344	14 ± 6	1702 ± 253	1155 ± 200	9190 ± 1473	115 ± 21	390 ± 184	3927 ± 1043	10	1.42 ± 0.03 a*
Zikhron Ya'akov	6 ± 3	17 ± 5	20 ± 3	2617 ± 491	20 ± 0	1507 ± 66	1167 ± 58	11220 ± 479	120 ± 10	520 ± 131	4630 ± 701	10	1.29 ± 0.12 ab
'Ofer	5 ± 0	15 ± 2	9 ± 4	2787 ± 232	21 ± 7	1277 ± 111	1200 ± 0	10050 ± 459	130 ± 10	387 ± 185	4217 ± 397	10	1.28 ± 0.07 ab
En HaShofet	8 ± 1	17 ± 0	27 ± 4	3830 ± 1339	26 ± 11	1393 ± 435	1467 ± 153	12267 ± 3653	129 ± 10	1193 ± 558	4710 ± 1991	9	1.26 ± 0.18 ab
Shemurat Allon	9 ± 2	18 ± 2	41 ± 9	3177 ± 657	25 ± 5	1447 ± 287	1367 ± 58	8583 ± 93	120 ± 10	587 ± 119	3373 ± 738	10	1.24 ± 0.15 ab
Or'Aquiva	9 ± 1	21 ± 1	16 ± 6	3143 ± 393	21 ± 8	1860 ± 20	1600 ± 0	10773 ± 665	173 ± 15	1480 ± 131	2747 ± 257	10	1.23 ± 0.08 ab
Jasar A-Zarka	10 ± 0**	18 ± 7**	28 ± 2**	2235 ± 219**	15 ± 5**	1235 ± 7**	1300 ± 283	14155 ± 615**	160 ± 14**	1295 ± 827**	4490 ± 141**	10	1.23 ± 0.11 ab
Nahsholim	6 ± 2	20 ± 1	25 ± 11	2630 ± 157	17 ± 10	1830 ± 195	1400 ± 264	14260 ± 1659	173 ± 6	743 ± 230	5893 ± 871	10	1.20 ± 0.22 ab
Pardes Hana	3 ± 1	18 ± 2	21 ± 5	2817 ± 244	30 ± 10	1123 ± 117	1667 ± 115	10056 ± 176	163 ± 6	303 ± 61	2177 ± 272	10	1.19 ± 0.16 ab
Ma'agan Mikhael	13 ± 1	25 ± 2	31 ± 5	3293 ± 604	17 ± 4	1673 ± 248	1333 ± 58	10197 ± 1319	118 ± 16	1380 ± 50	5060 ± 1647	10	1.19 ± 0.17 ab
Tel Aviv University (old Campus)	10 ± 1	15 ± 5	58 ± 11	2495 ± 431	41 ± 10	1198 ± 372	1333 ± 153	7967 ± 3409	147 ± 23	849 ± 772	2870 ± 1230	10	1.17 ± 0.10 bc
Kefar HaYarok Junction	14 ± 5	15 ± 7	158 ± 42	3823 ± 871	24 ± 7	1080 ± 108	1467 ± 153	9220 ± 1041	157 ± 25	467 ± 94	3563 ± 1003	10	1.15 ± 0.23 bc
Tel Aviv (southwest)	14 ± 1	23 ± 2	134 ± 17	2690 ± 131	57 ± 3	113 ± 121	2067 ± 153	9833 ± 812	207 ± 15	1130 ± 46	2943 ± 789	20	1.03 ± 0.22 c

Note: All data referring to the element content are in ppm/dry wt and are based on 3 to 4 replicates, except those marked by asterisks, which are based on duplicates. (*, values followed by the same letter do not differ significantly at $P = 0.05$ by Duncan's multiple range test; **, two replicates only

From Garty, J., Ronen, R., and Galun, M., *Environ. Exp. Bot.*, 25, 67, 1985. With permission.

Table 3
EFFECT OF SHORT-TERM EXPOSURES TO AQUEOUS SULFUR DIOXIDE ON POTASSIUM RETENTION BY THE LICHEN THALLUS

Exposure time (min)	Treatment	Amount of potassium leakage into the incubation medium (ppm)	Amount of potassium leakage into the incubation medium after an interim 24-hr recovery period (ppm)
15	Disks treated with distilled water (pH 5.8)	0.30	0.20
60		0.30	0.20
15	Disks treated with acidified distilled water (pH 3.0)	0.35	0.20
60		0.35	0.20
15	Disks treated with aqueous sulfur dioxide (75 ppm, pH 3.0)	1.40	0.45
60		2.60	4.60

From Puckett, K. J., Richardson, D. H. S., Flora, W. P., and Nieboer, E., *New Phytol.*, 73, 1183, 1974. With permission.

Experimentally, the effect of sulfur dioxide has been simulated by either gaseous SO_2 or treatment with sodium bisulfite or sodium sulfite.

Aqueous concentrations of SO_2 can be converted to gaseous values by the experimentally verified relationship:[76,77]

$$\text{ppm } SO_2 \text{ (wt/wt, aqueous)} = 10.3\sqrt{\text{ppm } SO_2 \text{ (v/v, in air)}}$$

The advantages and disadvantages of gaseous vs. aqueous forms of SO_2 for lichen sensitivity experiments have been pointed out by Richardson and Nieboer.[77] Generally, it is much more convenient to experiment with aqueous exposures, because it is much easier to prepare low concentrations of sulfite or bisulfite than low concentrations of gaseous SO_2, which has to be continuously monitored. Moreover, since dry thalli are not affected, measurements taken after aqueous SO_2 exposures seem to simulate the ''natural'' situation more realistically.

Mechanisms of SO_2 toxicity in lichens and higher plants have been summarized by Richardson and Nieboer[77] and are presented in Tables 4 and 5.

Why are physiological processes of sensitive lichen species by far more affected by atmospheric pollution than such processes in most vascular plants? It is at least partially due to the lack of any regulatory mechanism for water absorption or water loss in lichens. All nutrients and substances carried down by precipitation are taken up by the entire thallus. Water loss is also governed by the environment and the evaporated water leaves behind the toxic elements.

VI. UPTAKE AND ACCUMULATION OF ELEMENTS IN LICHENS

The mechanism(s) involved in the continuous (or long life) existence of the resistant lichen species, whether by avoidance or by specifically regulated tolerance mechanisms, still has to be critically studied. Yet, such species are very useful for monitoring the fallout of heavy metals and other elements.

Table 4
GENERAL MECHANISMS OF SO_2 TOXICITY

Type of reaction	Observed/expected response or injury
Enzyme deactivation Chemical modification (e.g., sulfitolysis) Binding to metal centers (vitamin B_{12}; Fe) Inhibition (competition with HCO_3^-; $H_2PO_4^-$)	Reduced metabolic activity; loss of membrane integrity, membrane function, and cell osmolality
Stimulation of enzyme systems (often in response to low pollution levels) Glucose-6-phosphate-dehydrogenase Increase in glutathione and total protein SH	Use of increases in enzyme activity as a bioindicator for nonvisible injury metabolism and detoxification of absorbed SO_2
Reaction with reactive biomolecules Chemical (bisulfite adducts) Redox (acts as electron acceptor/donor at pH 7)	Modification of metabolic precursors and products; interference with electron flow in photosynthetic and respiratory electron-transport chains

From Richardson, D. H. S. and Nieboer, E., *J. Hattori Bot. Lab.*, 54, 331, 1983. With permission.

A. Sulfur

The ability of lichens to accumulate high levels of sulfur has been demonstrated in many investigations.[78] Many of them suggest a positive correlation between the sulfur content in lichens and the amount of SO_2 in the atmosphere to which they were exposed.[21,77,79-82] Some studies, however, show that the amount of sulfur absorbed is species dependent, and not all species from the same site would contain the same amount of sulfur.[83] Such differences may be the result of different retention capacities. Sulfur content also has been measured in sensitive species and the amounts correlated to the impact on physiological parameters.[62,84]

B. Heavy Metals

Absorption and accumulation of heavy metals from the environment are well-documented features of lichens and have been used to monitor atmospheric depositions of a large number of metals. A linear relationship between the concentrations of several elements and the reciprocal of the distance from the emission source has been observed along transects near many industries and power plants.[19,86,89,90] A variety of elements are often accumulated simultaneously in one specimen.[12a,82,85-88]

Accumulation surveys have been recommended as baseline studies for periodic assessments where industrial activity is in progress and for developing urbanization areas.[12a,86] Content analyses can be carried out by X-ray fluorescence spectrometry, atomic absorption spectrophotometry, or neutron activation. The amounts of elements detected, some highly toxic, are large and often much higher than concentrations a living organism could possibly sustain (e.g., 3000 ppm lead,[91] 127 ppm chromium,[92] 335 ppm cadmium[93]).

Garty et al.[94] demonstrated the extracellular deposition of metallic particles ranging in size from 10 to 30 μm, and determined their elemental composition by means of energy dispersion X-ray analysis in combination with scanning electron microscopy (see Figure 3). The retention capacity of particulate contaminants embedded in the amorphous material of the relatively large interhyphal (mainly medullary) spaces apparently prevents solubilization and blocks an intrusion into the cells.

Table 5
TOXICITY MECHANISMS INVOLVING FORMATION OF RADICALS

Mode of generation	Observed/expected response or injury
Reduction of SO_2 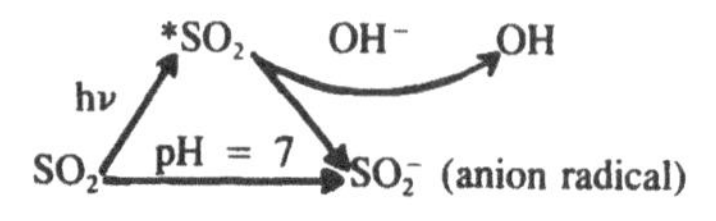 Vit C, NAD(P)H, $FADH_2$	Enhanced sulfitolysis; depletion of reducing agents such as NADPH, FADH, ascorbic acid, and glutathione; attack at electrophilic centers on enzymes and membranes
Production of active oxygen radicals SOD (superoxide dismutase) $\xrightarrow{HSO_3^-}$ inactive enzyme and O_2^- build up in chloroplasts $HSO_3^- + O_2^-$ (super oxide radical initiator) → 'active oxygen' (O_2^-, $^1O_2$, H_2O_2, OH·) Chlorophyll a + O_2^- → bleached products Membrane lipids $\xrightarrow{^1O_2}$ lipid hydroperoxides	Bleaching of chlorophyll; injury to chloroplast and other membranes (perhaps by degradation of unsaturated fatty acids); functional group destruction in biomolecules

From Richardson, D. H. S. and Nieboer, E., *J. Hattori Bot. Lab.*, 54, 331, 1983. With permission.

Another important accumulation mechanism is via extracellular binding or complexing of metal ions by an ion-exchange process, which occurs mainly on the hyphal cell walls.[10,78,95]

In relation to their physiological toxicity, the heavy metals in solution have been divided into three categories:[96] class A includes the alkaline and the alkaline-earth elements (e.g., K^+ and Sr^{2+}) which are relatively nontoxic and tend to form ligands with hydroxyl and other oxygen-containing sites on organic molecules; class B includes the heavy metals (e.g., Ag^+, Cu^{2+}, and Mg^{2+}) which may form ligands with nitrogen- and sulfur-containing groups; and class C is a borderline group which includes Zn^{2+}, Ni^{2+}, Pb^{2+}, and Cu^{2+} and may form ligands with either of the above. The authors remark that "The generally accepted term 'heavy metal' lacks rigorous definition and encompasses a heterogeneous array of elements. The proposed classification of metal ions separates them into three chemically and biologically significant categories."

C. Radionuclides

In addition to the extremely slow growth rate and long endurance of lichens, which result in their being very efficient collectors of air-borne contaminants over long periods of time, the microtopography of the thallus surface is an important factor in fallout interception rates. Lichen thalli have wrinkled, uneven, and porous surfaces (see Figures 1 to 9, 35, 51, and

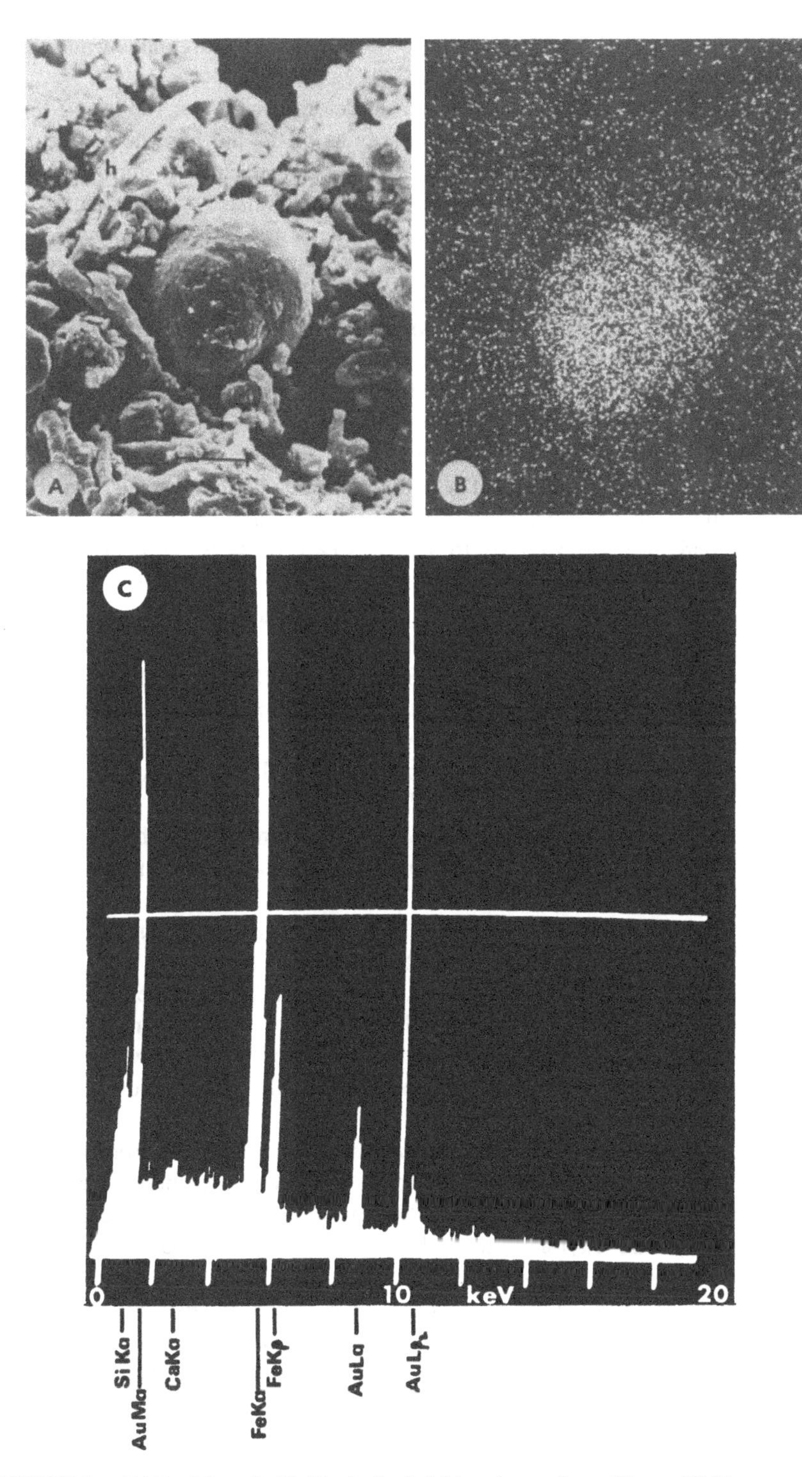

FIGURE 3. (A) Particle embedded in the hyphal (h) region; scale = 10 μm. (B) X-ray mapping of the identical sample area shown in (A) with the energy dispersive detector set at a window corresponding to the energy of FeKα (6240 eV). (C) The full energy spectrum from particle (A) obtained by using the multichannel analyzer. Besides the prominent Kα and Kβ peaks of Fe, the three characteristic peaks of gold due to the coating are noticeable and, in addition, there are peaks of Si and Ca. (From Garty, J., Galun, M., and Kessel, M., *New Phytol.*, 82, 159, 1979. With permission.)

84 in Volume 1; Chapter III) which certainly contribute to the entrappment and uptake of fall-out deposits including radionuclides.

Radionuclides derived from the fallout of nuclear detonations in the 1950s and 1960s had the greatest effect in the Arctic and sub-Arctic regions of the northern hemisphere, where lichens constitute the major source of food for reindeer and caribou during winter. The radionuclides deposited in lichens are transferred to the animals, whose meat is used for human consumption. During 1962 to 1964, body burdens of ^{137}Cs in Lapps were 30 to 40 times higher than in other Scandinavians.[97] The ^{137}Cs body burdens of the people were generally proportional to the consumption of reindeer meat, and the ^{137}Cs levels in the reindeer meat were related to their seasonal consumption of lichens. Lichens (mainly *Cladonia alpestris, C. rangiferina,* and *C. silvatica*) are the main fodder for a winter period of about 6 months. Therefore, the maximal ^{137}Cs burden was at the end of winter (March) and the minimum at autumn (September).[98,99] After the discovery of these high body burdens of ^{137}Cs in Swedish Lapps, more extensive analyses of the radionuclide content of lichens and other vegetation were carried out. The maximum area-content of ^{137}Cs in the lichen carpet (measured in *C. alpestris*) occurred in 1965 and 1966, and amounted to 46 to 66 nCi/m^2, which corresponds to a radioactivity to dry mass ratio of 34 to 52 nCi/kg.[100]

At first, most attention was devoted to ^{137}Cs (which exchanges, irreversibly, the calcium in human bones), because of its public health aspects and also for technical reasons.[98] In the 1970s, new γ-spectrometrical methods were introduced and other artificially produced emitting radionuclides (e.g., ^{238,239}Pu, ^{134}Cs, ^{90}Sr, ^{55}Fe, and ^{22}Na) were also measured.[101,102] All these radionuclides are also subjected to high retention in the lichen carpet, but each penetrates the carpet in its own special way. According to increasing depth penetration in the lichen carpet, the radionuclides may be arranged in the following sequence: $^{144}Ce \sim {}^{7}Be \ll {}^{95}Zr < {}^{137}Cs < {}^{106}Ru \lesssim {}^{155}Eu \ll {}^{125}Sb$.[102]

The vertical distribution of the radionuclides is of special interest and has to be measured from time to time, because the reindeer consume only the fresh (approximately 3 cm) upper layer of the lichen carpet.

The content of the ^{137}Cs was also measured in *X. parietina* and *P. saxatilis* from rocky shores of the Baltic Coast in order to evaluate the risk to sunbathers from exposure to radioactive wastes.[103] The seashore lichens contained significantly higher ^{137}Cs concentrations than the inland specimens, because in addition to direct deposition and bird guano, they were continuously exposed to the radionuclides dissolved in the sea water. Considering the turnover rate of ^{137}Cs, it was calculated that 50% of the contribution from the 1965 fallout will still remain after 12 years.[103]

Samples of lichens and the Arctic cushion plant *Dryas integrifolia* from the Canadian Arctic collected during the spring and summer of 1977 showed measurable amounts of the short-lived isotopes ^{144,141}Ce, ^{103}Ru, ^{95}Zr, and ^{95}Nb, in addition to previously deposited ^{137}Cs. According to Svoboda and Taylor,[104] these isotopes are clearly products of the Chinese nuclear explosion of September 1976.

A study[105] on *C. rangiferina* and *Stereocaulon* sp. in the Northwest Territories, Canada, the crash zone of Cosmos 945, revealed the presence of radioisotopes, but did not exceed the radioactivity contributed by the Chinese nuclear explosions of September 1976 and December 1978.

The recent emission of radionuclides from the Chernobyl atomic power plant (April 1986) has probably vastly contaminated the Arctic lichen carpets. This contamination is expected to impose a devastating effect on the reindeer and caribou herds and, consequently, on the health of the Lapps in Northern Scandinavia. The extent of this catastrophy should have its record in the measured levels of radionuclides in the Arctic lichens and should be compared to previous fallout events.

REFERENCES

1. **Nylander, W.**, Les lichens du Jardin du Luxembourg, *Bull. Soc. Bot. Fr., (Tome XIII), C. R. Seances*, 5, 364, 1866.
2. **Nylander, W.**, *Les Lichens des Environs de Paris*, 1896.
3. **Case, J. W.**, Lichen biomonitoring networks in Alberta, *Environ. Monitoring Assessment*, 4, 303, 1984.
4. **Grindon, L. H.**, *The Manchester Flora*, W. White, London, 1859.
5. **MacMillan, H.**, *Footnotes from the Page of Nature on First Forms of Vegetation*, Macmillan, London, 1861.
6. **Brodo, J. M.**, Lichen growth and cities: a study on Long Island, New York, *Bryologist*, 69, 427, 1966.
7. **Ferry, B. W., Baddeley, M. S., and Hawksworth, D. L., Eds.**, *Air Pollution and Lichens*, The Athlone Press, London, 1973.
8. **Sundström, K. R. and Hallgren, J. E.**, Using lichens as physiological indicators of sulfurous pollutants, *Ambio*, 2, 13, 1973.
9. **Deruelle, S.**, Les lichens et la pollution atmosphérique, *Bull. Ecol.*, 9, 87, 1978.
10. **Richardson, D. H. S. and Nieboer, E.**, Lichens and pollution monitoring, *Endeavour*, 5, 127, 1981.
11. **Hawksworth, D. L. and Rose, F.**, *Lichens as Pollution Monitors*, Stud. Biol. No. 66, Edward Arnold, London, 1976.
12. **Nash, T. H., III and Sigal, L.**, Ecological approaches to the use of lichenized fungi as indicators of air pollution, in *The Fungal Community: Its Organization and Role in the Ecosystem*, Wicklow, D. and Carrol, G., Eds., Marcel Dekker, New York, 1981.

12a. **Garty, J., Fuchs, C., Zisapel, N., and Galun, M.**, Heavy metals in the lichen *Caloplaca aurantia* from urban, suburban and rural regions in Israel (a comparative study), *Water, Air Soil Pollut.*, 8, 171, 1977.

13. **Sernander, R.**, Granskor och Fiby urskog, *Acta Phytogeogr. Suec.*, 8, 1, 1926.
14. **Skye, E.**, Lichens and air pollution, *Acta Phytogeogr. Suec.*, 52, 1, 1968.
15. **Lundstrom, H.**, The effect of air pollution on the epiphytic flora of conifers in the Stockholm region, *Stud. For. Suec.*, 56, 1, 1968.
16. **Seitz, W.**, Flechtenwuchs und Luftverunreinigung im Grossraum von Saarbrücken, *Ber. Dtsch. Bot. Ges.*, 85, 129, 1972.
17. **Johnsen, I. and Söchting, U.**, Influence of air pollution on the epiphytic lichen vegetation and bark properties of deciduous trees in the Copenhagen area, *Oikos*, 24, 344, 1973.
18. **Kurokawa, S.**, Preliminary studies on lichens of urban areas in Japan, in *Fundamental Studies in the Characteristics of Urban Ecosystems*, Numata, M., Ed., 1973, 80.
19. **Sugiyama, K., Kurokawa, S., and Okadam, G.**, Studies on lichens as a bioindicator of air pollution. Correlation of distribution of *Parmelia tinctorum* with SO_2 air pollution, *Jpn. J. Ecol.*, 26, 109, 1976.
20. **Züst, S.**, *Die Epiphytenvegetation in Raume Zürich als Indikator der Umweltbelastung*, Heft 62, Veröffentlichungen des Geobotanischen Institutes der Eidg. Techn. Hochschule, Stiftung Rubel, Zürich, 1977.
21. **Hopp, U. and Kappen, L.**, Einige Aspekte zur Immissionbedingten Verbreitung von Flechten im Stadtgebiet von Würzburg, *Ber. Bayer. Bot. Ges.*, 52, 15, 1981.
22. **Schmid, A. B.**, Die Epixyle Flechtenvegetation von München, thesis, Naturw. Fak. Univ., München, 1957.
23. **Rose, C. I. and Hawksworth, D. L.**, Lichen recolonization in London's cleaner air, *Nature (London)*, 289, 289, 1981.
24. **Henderson-Sellers, A. and Seaward, M. R. D.**, Monitoring lichen reinvasion of ameliorating environments, *Environ. Pollut.*, 19, 207, 1979.
25. **Barkman, J. J.**, *Photosociology and Ecology of Cryptogamic Epiphytes*, Van Gorcum & Co., Assen, 1958.
26. **Rydzack, J.**, Influence of small towns on the lichen vegetation. VII. Discussion and general conclusions, *Ann. Univ. Marie-Curie Sklodowska*, 13C, 275, 1959.
27. **Coppins, B. J.**, The drought hypothesis, in *Air Pollution and Lichens*, Ferry, B. W., Baddeley, M. S., and Hawksworth, D. L., Eds., The Athlone Press, London, 1973, 124.
28. **LeBlanc, F. and Rao, D. N.**, Evaluation of the pollution and drought hypothesis in relation to lichens and bryophytes in urban environments, *Bryologist*, 76, 1, 1973.
29. **Trass, H.**, An index for the utilization of lichen groups to determine air pollution, *Eesti Loodus*, 11, 268, 1968.
30. **LeBlanc, F. and De Sloover, J.**, Relation between industrialization and the distribution and growth of epiphytic lichens and mosses in Montreal, *Can. J. Bot.*, 48, 1485, 1970.
31. **De Sloover, J.**, Pollutions atmosphériques et tolérance spécifique chez les lichens, *Soc. Bot. Fr. Colloq. Lich.*, 205, 1967.
32. **LeBlanc, F., Rao, D. N., and Comeau, G.**, The epiphytic vegetation of *Populus balsamifera* and its significance as an air pollution indicator in Sudbury, Ontario, *Can. J. Bot.*, 50, 519, 1972.

33. **Crespo, A., Manrique, E., Barreno, E., and Serina, E.,** Valoracion de la contaminacion atmosferica del area urbana de Madrid mediante bioindicadores (Liquenes epifitos), *Anal. Inst. Bot. A. J. Cavanilles,* 34, 71, 1977.
34. **Deruelle, S.,** Etude comparée de la sensibilite de trois méthodes d'estimation de la pollution atmosphfique, en utilisant les lichens comme indicateurs biologiques, dans la region de Mantes (Yvelines), *Rev. Bryol. Lichenol.,* 44, 429, 1978b.
35. **Hawsworth, D. L. and Rose, F.,** Qualitative scale for estimating sulphur dioxide air pollution in England and Wales using epiphytic lichens, *Nature (London),* 227, 145, 1970.
36. **Deruelle, S.,** Influence de la pollution atmosphérique sur la végétation lichénique des arbres isolés dans la région de Mantes (Yvelines), *Rev. Bryol. Lichenol.,* 43, 35, 1977.
37. **Gilbert, O. L.,** A biological scale for the estimation of sulphur dioxide pollution, *New Phytol.,* 69, 629, 1970.
38. **Trass, H.,** Lichen sensitivity to the air pollution and index of poleotolerance (I.P.), *Fol. Crypt. Est.,* 3, 19, 1973.
39. **Brodo, I. M.,** Transport experiments with corticolous lichens using a new technique, *Ecology,* 42, 838, 1961.
40. **Schönbeck, H.,** A method for determining the biological effects of air pollution by transplanted lichens, *Staub. Reinhalt. Luft,* 29, 17, 1969.
41. **Wood, C. W., Jr. and Nash, T. H., III,** Copper smelter effluences effects on Sonoran desert vegetation, *Ecology,* 57, 1311, 1976.
42. **Dawson, J. L. and Nash, T. H., III,** Effects of air pollution from copper smelters on a desert grassland community, *Environ. Exp. Bot.,* 20, 61, 1980.
43. **Eversman, E.,** Lichens as predictors and indicators of air pollution from coal-fired power-plant emissions, in *The Bioenvironmental Impact of a Coal-Fired Power-Plant,* Colstrip, Mont., 1976, 91.
44. **Marsh, J. E. and Nash, T. H., III,** Lichens in relation to the Four Corners Power Plant in New Mexico, *Bryologist,* 82, 20, 1979.
45. **Nash, T. H., III and Sommerfeld, M. R.,** Elemental concentrations in lichens in the area of the Four Corners Power Plant, New Mexico, *Environ. Exp. Bot.,* 21, 158, 1981.
46. **Morgan-Huws, D. I. and Haynes, F. N.,** Distribution of some epiphytic lichens around an oil refinery at Fawley, Hampshire, in *Air Pollution and Lichens,* Ferry, B. W., Baddeley, M. S., and Hawksworth, D. L., Eds., The Athlone Press, London, 1973, 89.
47. **Nash, T. H., III,** Influences of effluents from a zinc factory on lichens, *Ecol. Monogr.,* 45, 183, 1975.
48. **Pyatt, F. B.,** Lichens as indicators of air pollution in a steel producing town in South Wales, *Environ. Pollut.,* 1, 45, 1970.
49. **Holopainen, T. H.,** Cellular injuries in epiphytic lichens transplanted to air polluted areas, *Nord. J. Bot.,* 4, 393, 1984.
50. **LeBlanc, F., Cameau, G., and Rao, D. N.,** Fluoride injury symptoms in epiphytic lichens and mosses, *Can. J. Bot.,* 49, 1691, 1971.
51. **Gilbert, O. L.,** The effect of SO_2 on lichens and bryophytes around Newcastle upon Tyne, in *Air Pollution,* Proc. 1st Eur. Congr. Influence of Air Pollution on Plants and Animals, Centre Agr. Publ. Docum., Wageningen, 1968, 233.
52. **Kauppi, M.,** Fluorescence microscopy and microfluorometry for the examination of pollution damage in lichens, *Ann. Bot. Fenn.,* 17, 163, 1980.
53. **LeBlanc, F. and Rao, D. N.,** Effects of sulphur dioxide on lichen and moss transplants, *Ecology,* 54, 612, 1973.
54. **Slocum, R. D.,** Effects of SO_2 and pH on the Ultrastructure of the *Trebouxia* Phycobiont of the Pollution-Sensitive *Parmelia caperata* (L.) (Ach.), M.Sc. thesis, Ohio State University, Columbus, 1977.
55. **Türk, R.,** Zur SO_2-Resistenz von Flechten Verschiedener Wuchsform, *Flora,* 164, 133, 1975.
56. **Türk, R., Wirth, V., and Lange, O. L.,** CO_2-Gaswechsel-Untersuchungen zur SO_2-Resistenz von Flechten, *Oecologia,* 15, 33, 1974.
57. **Puckett, K. J., Nieboer, E., Flora, W. P., and Richardson, D. H. S.,** Sulphur dioxide: its effects on photosynthetic ^{14}C fixation in lichens and suggested mechanisms of phototoxicity, *New Phytol.,* 72, 141, 1973.
58. **Ronen, R., Canaani, O., Garty, J., Cahen, D., Malkin, S., and Galun, M.,** Photosynthetic parameters in *Ramalina duriaei,* in vivo, studied by photoacoustics, in *Lichen Physiology and Cell Biology,* Brown, D. H., Ed., Plenum Press, New York, 1985, 9.
59. **Ronen, R., Canaani, O., Garty, J., Cahen, D., Malkin, S., and Galun, M.,** The effect of air pollution and bisulfite treatment in the lichen *Ramalina duriaei* studied by photoacoustics, in *Advances in Photosynthesis Research,* Vol. 4, Proc. 6th Congr. Photosynthesis, Brussels, Sybesma, C., Ed., Martin Nijhoff, The Hague, 1984.
60. **Lange, O. L.,** Der CO_2-Gaswechsel von Flechten nach Erwarmung im Feuditen Zustand, *Ber. Dtsch. Bot. Ges.,* 78, 441, 1965.

61. **Kappen, L. and Lange, O. L.,** Die Kalteresistenz einiger Makroflechten, *Flora*, 161, 1, 1972.
62. **Malhotra, S. S. and Khan, A. A.,** Sensitivity to SO_2 of various metabolic processes in an epiphytic lichen, *Evernia mesomorpha*, *Biochem. Physiol. Pflanz.*, 178, 121, 1983.
63. **Ewald, D. and Schlee, D.,** Biochemical effects of sulphur dioxide on proline metabolism in the alga *Trebouxia* sp., *New Phytol.*, 94, 235, 1983.
64. **Canaani, O., Ronen, R., Garty, J., Cahen, D., Malkin, S., and Galun, M.,** Photoacoustic study of the green alga *Trebouxia* in the lichen *Ramalina duriaei in vivo*, *Photosynthesis Res.*, 5, 297, 1984.
65. **Canaani, O., Ronen, R., Garty, J., Cahen, D., Malkin, S., and Galun, M.,** *Photoacoustic Study of Photosynthesis in Lichens*, 4th Int. Top. Meet. Photoacoustic, Thermal, and Related Sciences, L'Esterel, Quebec, 1985.
66. **Ronen, R.,** The Effect of Air Pollution on Physiological Parameters of the Lichen *Ramalina duriaei*, Ph.D. thesis, Tel Aviv University, Tel Aviv, 1986.
67. **Hallgren, J. E., and Gezelius, K.,** Effects of SO_2 on photosynthesis and ribulose biphosphate carboxylase in pine tree seedlings, *Physiol. Plant.*, 54, 153, 1982.
68. **Beekley, P. K. and Hoffman, G. R.,** Effects of sulfur dioxide fumigation on photosynthesis, respiration and chlorophyll content of selected lichens, *Bryologist*, 84, 379, 1981.
69. **Brown, D. H. and Hooker, T. N.,** The significance of acidic lichen substances in the estimation of chlorophyll and phaeophytin in lichens, *New Phytol.*, 78, 617, 1977.
70. **Ronen, R. and Galun, M.,** Pigment extraction from lichens with dimethyl sulfoxide (DMSO) and estimation of chlorophyll degradation, *Environ. Exp. Bot.*, 24, 239, 1984.
71. **Puckett, K. J., Richardson, D. H. S., Flora, W. P., and Nieboer, E.,** Photosynthetic ^{14}C fixation by the lichen *Umbilicaria mühlenbergii* (Ach.) Tuck. following short exposures to aqueous sulphur dioxide, *New Phytol.*, 73, 1183, 1974.
72. **Hallgren, J. E. and Huss, K.,** Effects of SO_2 on photosynthesis and nitrogen fixation, *Physiol. Plant.*, 34, 171, 1975.
73. **Sheridan, R. P.,** Impact of emissions from coal-fired electricity generating facilities on N_2-fixing lichens, *Bryologist*, 82, 59, 1979.
74. **Tomassini, F. D., Lavoie, P., Puckett, K. J., Nieboer, E., and Richardson, D. H. S.,** The effect of time of exposure to sulphur dioxide on potassium loss from and photosynthesis in the lichen *Cladina rangiferina* (L.) Harm., *New Phytol.*, 79, 147, 1977.
75. **Pearson, L. C.,** Air pollution damage to cell membranes in lichens, *Atmos. Environ.*, 19, 209, 1985.
76. **Puckett, K. G., Tomassini, F. D., Nieboer, E., and Richardson, D. H. S.,** Potassium efflux by lichen thalli following exposure to aqueous sulphur dioxide, *New Phytol.*, 79, 135, 1977.
77. **Richardson, D. H. S. and Nieboer, E.,** Ecological responses of lichens to sulphur dioxide, *J. Hattori Bot. Lab.*, 54, 331, 1983.
78. **Nieboer, E. and Richardson, D. H. S.,** Lichens as monitors of atmospheric deposition, in *Atmospheric Pollutants in Natural Waters*, Eisenreich, S. J., Ed., Ann Arbor Science, Michigan, 1981, 339.
79. **Gilbert, O. L.,** Lichens and air pollution, in *The Lichens*, Ahmadjian, V. and Hale, M. E., Eds., Academic Press, New York, 1973.
80. **Laaksovirta, K. and Olkkonen, H.,** Epiphytic lichen vegetation and element contents of *Hypogymnia physodes* and pine needles examined as indicators of air pollution at Kokkola, West Finland, *Ann. Bot. Fenn.*, 14, 112, 1977.
81. **Pakarinen, P.,** Regional variations of sulphur concentrations in *Sphagnum* mosses and *Cladonia* lichens in Finnish bogs, *Ann. Bot. Fenn.*, 18, 275, 1981.
82. **Puckett, K. J. and Finegan, E. J.,** An analysis of the element content of lichens from the Northwest Territories, Canada, *Can. J. Bot.*, 58, 2073, 1980.
83. **Takala, K., Olkkonen, H., Ikonen, J., Jääskelainen, J., and Puumalainen, P.,** Total sulphur contents of epiphytic and terricolous lichens in Finland, *Ann. Bot. Fenn.*, 22, 91, 1985.
84. **Garty, J., Ronen, R., and Galun, M.,** Correlation between chlorophyll degradation and the amount of some elements in the lichen *Ramalina duriaei* (De Not.) Jatta, *Environ. Exp. Bot.*, 25, 67, 1985.
85. **Seaward, M. R. D., Bylinska, E. A., and Goyal, R.,** Heavy metal content of *Umbilicaria* species from the Sudety region of S.W. Poland, *Oikos*, 36, 107, 1981.
86. **Tomassini, F. D., Puckett, K. J., Nieboer, E., Richardson, D. H. S., and Grace, B.,** Determination of copper, iron, nickel and sulfur by X-ray fluorescence in lichens from the Mackenzie Valley, Northwest Territories, and the Sudbury district, Ontario, *Can. J. Bot.*, 54, 1591, 1976.
87. **Fuchs, C. and Garty, J.,** Elemental content in the lichen *Ramalina duriaei* (De Not.) Jatta at air quality biomonitoring stations, *Environ. Exp. Bot.*, 23, 29, 1983.
88. **Garty, J. and Fuchs, C.,** Heavy metals in the lichen *Ramalina duriaei* transplanted in biomonitoring stations, *Water Air Soil Pollut.*, 17, 175, 1982.
89. **Nieboer, E., Ahmed, H. M., Puckett, K. J., and Richardson, D. H. S.,** Heavy metal content of lichens in relation to distance from a nickel smelter in Sudbury, Ontario, *Lichenologist*, 5, 292, 1972.

90. **Tuominen, Y. and Jaakkola, T.,** Absorption and accumulation of mineral elements and radioactive nuclides, in *The Lichens,* Ahmadjian, V. and Hale, M. E., Eds., Academic Press, New York, 1973, 185.
91. **Seaward, M. R. D.,** Some observations on heavy metal toxicity and tolerance in lichens, *Lichenologist,* 6, 158, 1974.
92. **Seaward, M. R. D.,** Lichen ecology of the Scunthorpe heathlands. I. Mineral accumulation, *Lichenologist,* 5, 423, 1973.
93. **Nash, T. H., III,** Effect of Effluents from a Zinc Factory on Lichens, Ph.D. thesis, Rutgers University, New Brunswick, N.J., 1971.
94. **Garty, J., Galun, M., and Kessel, M.,** Localization of heavy metals and other elements accumulated in the lichen thallus, *New Phytol.,* 82, 159, 1979.
95. **Nieboer, E., Richardson, D. H. S., and Tomassini, F. D.,** Mineral uptake and release by lichens: an overview, *Bryologist,* 81, 226, 1978.
96. **Nieboer, E., Richardson, D. H. S., Lavoie, P., and Padovan, D.,** The role of metal-ion binding in modifying the toxic effects of sulphur dioxide on the lichen *Umbilicaria mühlenbergii.* I. Potassium efflux studies, *New Phytol.,* 82, 621, 1979.
97. **Hasanen, E. and Miettinen, J. K.,** Gamma-emitting radionuclides in subarctic vegetation during 1962 to 1964, *Nature (London),* 212, 379, 1966.
98. **Hanson, W. C.,** Cesium-137 in Alaskan lichens, caribou and eskimos, *Health Phys.,* 13, 383, 1967.
99. **Linden, K. and Gustafsson, M.,** Relationships and seasonal variation of ^{137}Cs in lichen, reindeer and man in northern Sweden, 1961 to 1965, in *Radioecological Concentration Processes,* Proc. Int. Symp., Stockholm, 1966, 193.
100. **Sören Mattsson, L. J.,** ^{137}Cs in the reindeer lichen *Cladonia alpestris;* deposition, retention and internal distribution, 1961 to 1970, *Health Phys.,* 28, 1975.
101. **Sören Mattsson, L. J.,** Deposition, retention and internal distribution of ^{155}Eu, ^{144}Ce, ^{125}Sb, ^{106}Ru, ^{95}Zr, ^{54}Mn and ^{7}Be in the reindeer lichen *Cladonia alpestris,* 1961—1970, *Health Phys.,* 29, 27, 1975.
102. **Holm, E. and Persson, R. B. R.,** Fall-out plutonium in Swedish reindeer lichens, *Health Phys.,* 29, 43, 1975.
103. **Larsson, J. E.,** ^{137}Cs in lichen communities on the Baltic coast, *Sven. Bot. Tidskr.,* 64, 173, 1970.
104. **Svoboda, J. and Taylor, H. W.,** Persistence and ^{137}Cs in Arctic lichens, *Dryas integrifolia* and lake sediments, *Arct. Alp. Res.,* 11, 95, 1979.
105. **Taylor, H. W., Hutchinson, E. A., McInnes, K. L., and Svoboda, J.,** Cosmos 954: search for airborne radioactivity on lichens in the crash area Northwest Territories, Canada, *Science,* 205, 1383, 1979.
106. **Silberstein, L., Keller, P., and Galun, M.,** unpublished results.
107. **Silberstein, L., and Galun, M.,** Spectrometric estimation of chlorophyll in lichens containing anthraquinones in relation to air pollution assessment, *Exp. Env. Bot.,* in press.

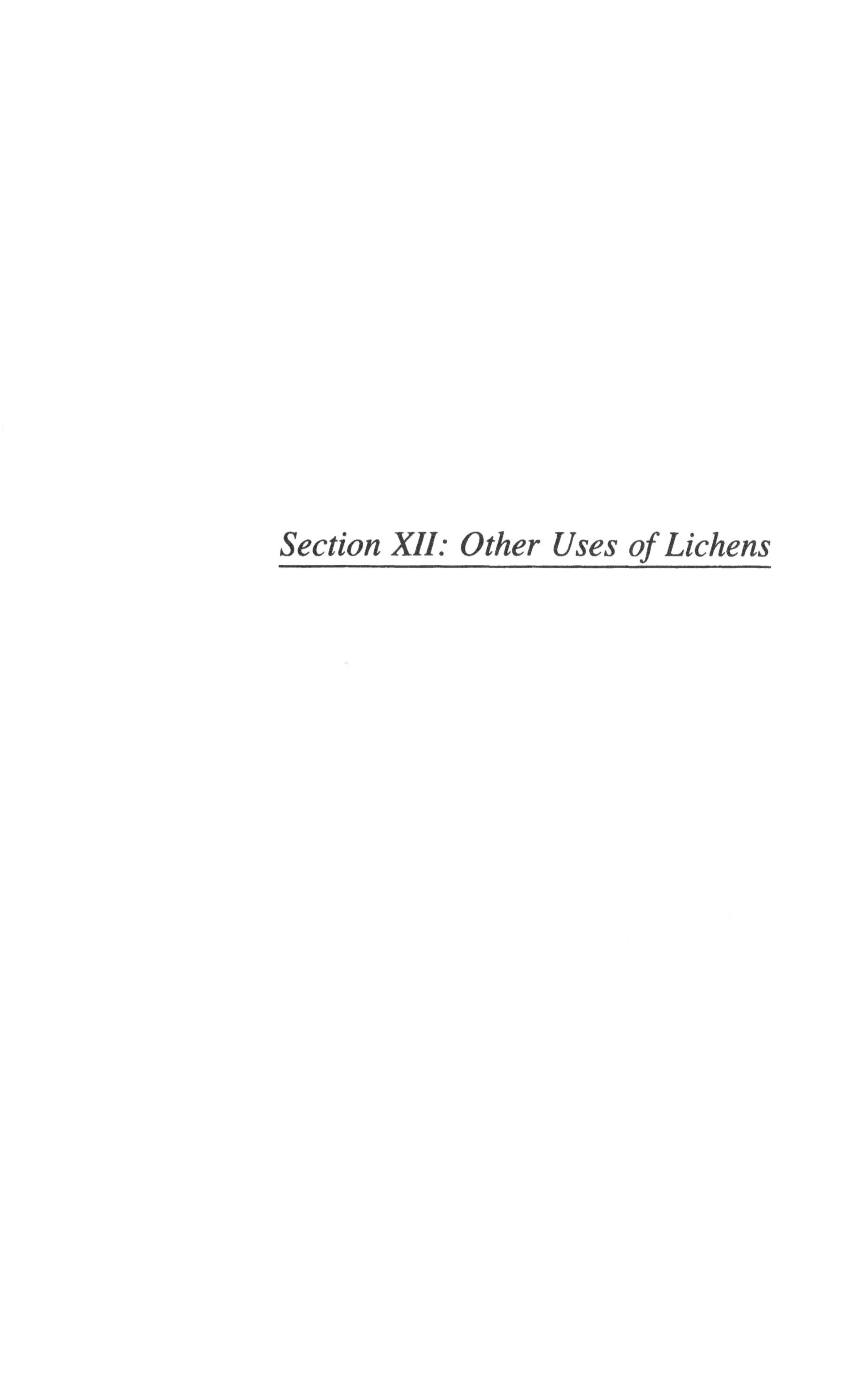

Section XII: Other Uses of Lichens

Chapter XII. A

THE USE OF LICHENS IN DATING

John L. Innes

I. INTRODUCTION

The basic principle underlying the use of lichens in dating (lichenometry) is that if the relationship between the size and age of a given lichen taxon is known, then the age of a surface can be inferred from the size of the lichens present on it. The technique has been applied widely within a variety of disciplines, although its greatest use has been in the fields of geomorphology and archaeology. It is considered to be particularly useful for dating Neoglacial deposits in arctic and alpine environments. This is because suitable materials for alternative dating techniques, such as ^{14}C assay, thermoluminescence, dendrochronology, and amino acid racemization, may not be present. Lichenometry is most useful for dating surfaces exposed within the last 500 years, as many to the alternative means for dating deposits do not cover this period. However, considerable difficulties surround the methodology of the technique and these are only just beginning to be resolved. Further work is still required, particularly on the ecology of target taxa. The aim of this account is to detail the main procedures involved in lichenometry and to identify some of the difficulties that still remain. A detailed review of the technique is given in Innes.[1]

II. TAXA USED

A wide variety of taxa have been used in dating studies. These are summarized in Table 1. The list has been drawn from the literature, and the accuracy of the identifications is unknown.

The majority of lichenometric studies deal with taxa within the *Rhizocarpon* genus (Figure 1). This is a composite group consisting of many taxa (Table 2). Whether these represent "species" as such is irrelevant. Several studies have shown that different taxa within the genus grow at different rates and that each may have specific environmental requirements.[1,44,55,56] The recognition of individual taxa in the field is frequently very difficult, although it may be possible to distinguish certain groups. In the past, there has been a tendency to term all thalli within the *Rhizocarpon* genus as *R. geographicum*, although some studies have acknowledged the composite nature of the group by using terms such as *R. geographicum* s.l.,[16,36,57-59] *Rhizocarpon geographicum* agg.,[40,60,62,63] *R. geographicum* sp.,[64] or *R. geographicum* coll.[43,64-66]

The potential effects of aggregating the taxa have been demonstrated by Innes.[1,55,56] *Rhizocarpon* section *Rhizocarpon* taxa can usually be separated from section *Alpicola* taxa in the field.[4,35,55,56,67] Further separation may not be possible without laboratory determinations. The two units grow at different rates. A relative growth rate model for southwest Norway has been presented by Innes[55] and is shown in Figure 2. *R. alpicola* colonizes surfaces after *Rhizocarpon* section *Rhizocarpon* taxa, but then grows faster, "overtaking" *Rhizocarpon* section *Rhizocarpon* taxa on surfaces of about 200 to 250 years in age. This relationship is not universal, and in coastal situations in north Norway, *R. alpicola* appears to grow more slowly than *Rhizocarpon* section *Rhizocarpon* taxa.[1,41] These variations may reflect the fairly strict environmental requirements of *R. alpicola*, but no quantitative studies have been made to confirm this. Similar variations in the relative growth rates of taxa within the *Rhizocarpon* genus have been found for *R. jemtlandicum* and section *Rhizocarpon* taxa in west Greenland.[30]

Table 1
TAXA USED OR RECOMMENDED FOR USE IN LICHENOMETRY

Taxa	Ref.[b]
Acarospora chlorophana (Wahlenb.) Mass.	3
Caloplaca cinericola (Hue) Darb.	4
Diploschistes anactinus (Nyl.) Zahlbr.	5
D. scruposus (Schreb.) Norm.	6
Huilia tuberculosa[a]	7
Lecanora alpina[a]	8
L. arctica[a]	9
L. aspicilia[a]	10
L. badia (Hoffm.) Ach.	6, 11
L. caesiocinerea Nyl. ex Malbr.	6, 12
L. calcarea (L.) Sommerf.	7, 13
L. campestris (Schaer.) Hue	7
L. cinerea (L.) Sommerf.	6, 11, 12, 14—16
L. muralis (Schreb.) Rabenh.	7
L. polytropa (Hoffm.) Rabenh.	17
L. thomsonii H.Magn.	10, 18—23
Lecidea atrobrunnea (Ram.) Schaer.	3, 10, 18, 20—22, 24, 25
L. fumida[a]	9
L. lapicida (Ach.) Ach.	6, 9, 10, 12, 17
L. pantherina (Hoffm.) Th.Fr.	26
L. paschalis Zahlbr.	5
L. promiscens Nyl.	6, 11, 12
L. silacea (Ach.) Ach.	27
Parmelia conspersa (Ach.) Ach.	28
P. glabratula (Nyl.) Nyl.	7
Physcia caesia (Hoffm.) Hampe	17, 29
P. dubia (Hoffm.) Lynge	6
P. picta (Swans.) Nyl.	5
P. teretiuscula (Ach.) Lynge	6, 11
Placynthium nigrum (Huds.) S. Gray	7
Pseudephebe minuscula (Nyl. ex Arnold) Brodo et Hawksw.	24, 30—37
P. pubescens (L.) Choisy	17, 29, 30, 35, 36
Rhizocarpon alpicola (Hepp.) Rabenh.	4, 6, 12, 27, 38—44
R. candidum Dodge	45, 46
R. eupetraeoides (Nyl.) Blomb et Forss.	35, 36
R. geographicum (L.) DC	many studies
R. inarense Vainio	35, 36
R. jemtlandicum Malme	24, 30, 32, 37
R. lecanorinum Anders	3
R. macrosporum Ras.	25, 47
R. superficiale (Schaer.) Vain.	3
Sporastatia testudinea (Ach.) Mass.	3, 12, 48, 49
Stereocaulon alpinum Laur.	17
Umbilicaria arctica (Ach.) Nyl.	17
U. cylindrica (L.) Del. ex Duby	6, 12, 43, 50, 51
U. hyperborea (Ach.) Hoffm.	17, 29, 30
U. proboscidea (L.) Schrad.	51
U. virginis Schaer.	17, 29
Usnea antarctica Du Rietz	52
Verrucaria muralis Ach.	7
V. nigrescens Pers.	7
Xanthoria elegans (Link.) Th.Fr.	17, 20, 29, 48, 53
X. parietina (L.) Th.Fr.	7

Table 1 (continued)
TAXA USED OR RECOMMENDED FOR USE IN LICHENOMETRY

[a] authors omitted in the original publication.
[b] Examples of usage are given in literature referenced.

Partly based on Locke, W. W., III, Andrews, J. T., and Webber, P. J., *Br. Geomorphol. Res. Group Tech. Bull.*, 26, 1, 1979. With permission.

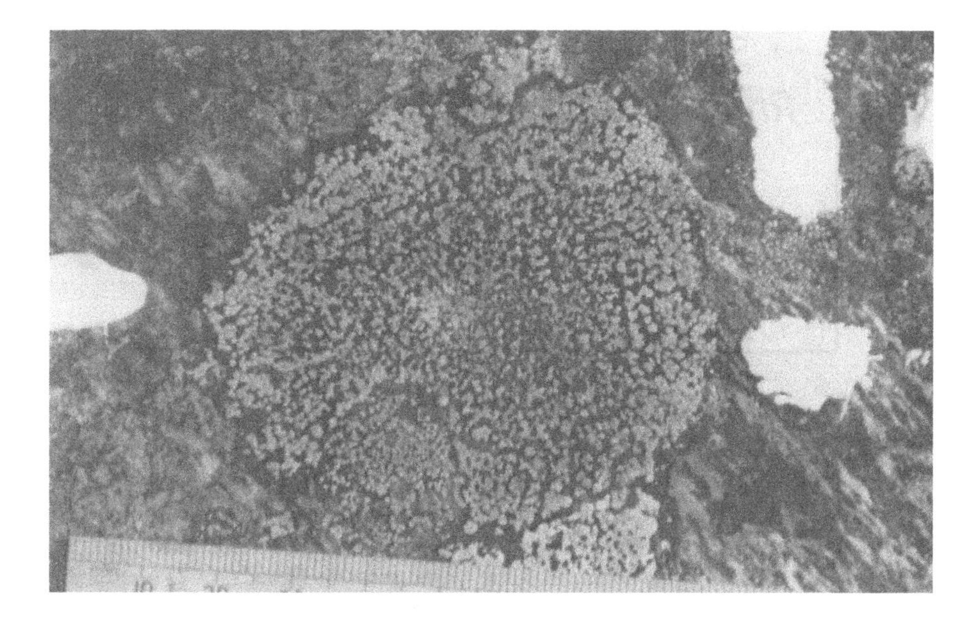

FIGURE 1. *Rhizocarpon geographicum* (L.) DC. This, and closely related taxa, is the most frequently used in lichenometry. Note the smaller thalli included at the base of the large thallus.

Table 2
CLASSIFICATION OF THE *RHIZOCARPON* GENUS[54]

***Rhizocarpon* genus**
Subgenus *Phaeothallus* — not normally used in lichenometry
Subgenus *Rhizocarpon*
 Section *Superficiale* (Runem.) Thoms.
 R. dispersum Runem., *R. effiguratum* (Anzi) Th.Fr., *R. norvegicum* Ras., *R. parvum* Runem., *R. pusillum* Runem., *R. superficiale* (Schaer.) Vain.
Section *Alpicola* (Runem.) Thoms.
 R. alpicola (Hepp.) Rabenh., *R. eupatraeoides* (Nyl.) Blomb. et Foss., *R. inarense* (Vain.) Vain.
Section *Viridiatrum* (Runem.) Thoms.
 R. atrovirellum (Nyl.) Zahlbr., *R. kakurgon* Poelt, *R. lusitanicum* (Nyl.) Arnold, *R. oportense* (Vain.) Ras., *R. subtile* Runem., *R. tetrasporum* Runem., *R. viridiatrum* (Wulf.) Koerb.
Section *Rhizocarpon* Cern.
 R. atroflavescens Lynge, *R. carpaticum* Runem., *R. ferax* H. Magn., *R. furax* Poelt et V. Wirth, *R. geographicum* (L.) DC., *R. intermediellum* Ras., *R. lecanorium* Anders, *R. macrosporum* Ras., *R. pulverulentum* (Schaer.) Ras., *R. rapax* V. Wirth et Poelt, *R. riparium* Ras., *R. saaense* Ras., *R. sphaerosporum* Ras., *R. sublucidum* Ras., *R. tavaresii* Ras., *R. tinei* (Tornab.) Runem. (s.str.)

Note: Only European taxa have been listed.

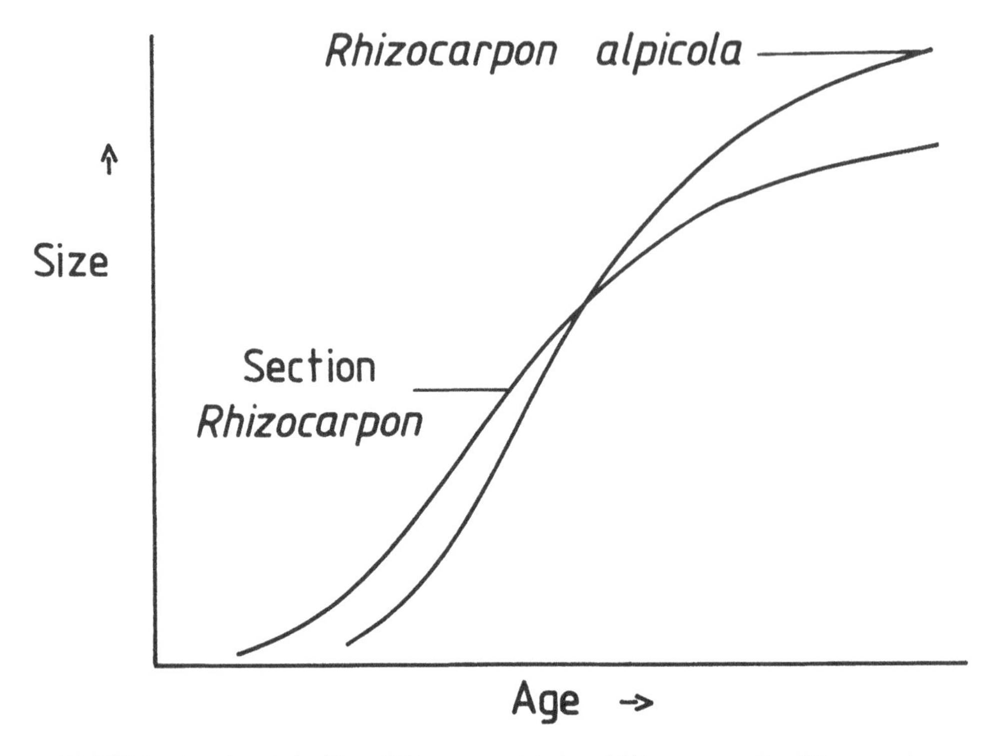

FIGURE 2. Age-size relationships of *Rhizocarpon alpicola* and *Rhizocarpon* section *Rhizocarpon* taxa.

No attempt has been made to separate the age-size relationships of taxa within the section *Rhizocarpon*, although these are also likely to vary. For example, *R. macrosporum* is known to grow faster than *R. geographicum* in some alpine areas of Canada.[25,47] This remains an area where major contributions could be made in the future.

The potential error in dating involved in aggregation is discussed by Innes.[55,56] Basically, lichenometric dating methods assume that the size-age relationship of lichens on one deposit can be transferred to another deposit. If the largest lichens on each deposit consist of different taxa (which is quite possible, due to environmental variations), then the assumption of identical size-age relationships may be violated. As a result, some lichenometric dates may be erroneous.

Further difficulties are associated with the division of the *Rhizocarpon* genus. In the literature, there are frequent references to "growth rates" of "*Rhizocarpon geographicum*". There are substantial difficulties in comparing these due to differences in sampling and measurement techniques (see below) and also because they frequently refer to different taxa.[68] To enable standardization, workers must use a strictly defined classification. Since one already exists within the taxonomic literature (Table 2), it is strongly recommended that workers dealing with *Rhizocarpon* taxa use it.

The taxa listed in Table 1 illustrate the variety of taxa that can be used in lichenometry. *Rhizocarpon* has been used in many studies as it is abundant in arctic-alpine environments and is long lived. The criteria for the use of a lichen taxon for dating follow:

1. It should be relatively common on all the surfaces under investigation.
2. It should have clearly defined margins.
3. The growth rate should be compatible with the ages of the surfaces.

The following account deals principally with studies using *Rhizocarpon*. However, it should be stressed that there are considerable advantages to be gained from using more than one taxon or group of taxa in any study.[7]

III. FIELD TECHNIQUES

Various problems surround the measurement of lichens for lichenometric purposes. The problems can be divided into those associated with measurement and those associated with sampling.

A. Measurement

There are several indexes that can be used to characterize the size of a lichen thallus. These include the longest axis, mean axis, shortest axis (= largest inscribed circle), and the area.[69] Both the mean axis and the area are extremely difficult to measure objectively, especially in irregular thalli. Consequently, few workers have used these two indexes. The largest inscribed circle has been used by a number of workers.[10,28,47,59,61,70] Locke et al.[2] argue that the use of this index reduces the possibility of including coalesced thalli and thereby obtaining an inflated value. However, with care, this potential source of error can be excluded by close examination of thalli. The following criteria can be used to identify composite thalli:

1. The presence of a continuous strip of prothallus running across the thallus
2. A change in the color of the areolae
3. Very irregular shapes where there is no evidence of factors that might have produced the irregularities

These are all subjective methods and there is at present no objective means of identifying composite thalli. If in doubt, the best policy is to exclude dubious thalli. Statistically, the subjective element in the measurement of the largest inscribed circles means that they are significantly less reproducible than longest axis measurements.[1] In view of the difficulties associated with the largest inscribed circle diameter, I recommend that workers use the longest axis of thalli as an index of size. This index is already used by the majority of workers.

Several studies have included only circular or near-circular thalli.[10,36,67,71,72] The rationale for such a strategy is again the avoidance of coalesced thalli. However, as a thallus grows, there will be an increasing probability that growth along one or more axes will be inhibited. This will result in a departure from circularity. Consequently, the total population of circular thalli will decrease through time. If only circular thalli are measured, the probability of determining the maximum possible size-age relationship will decline through time. As a result, it can be recommended that both circular and irregular thalli are included. There is no objective measure of how irregular a thallus can become. Birkeland[73] recommended a limiting ratio (minimum/maximum diameter) of 0.75, but there is no empirical justification for this, and many genuine thalli lie outside it.[1]

B. Number of Thalli

Several workers have argued that only the single largest thallus on a surface should be used for dating,[74] and this practice has been adopted in a number of recent studies.[16,66,70,75] The alternative approach is to adopt an averaging procedure.[26,60] The problem has been examined in detail by Innes.[76,77] The use of an averaging procedure reduces the potential effects of one or two anomalous thalli in a sample. The optimum sample size, in terms of the relationship between the search times and the reproducibility, appears to be the mean of the five largest thalli on a deposit. An important point to make is that size-age relationships derived for the mean of a given number of thalli must only be used for that number of thalli. This is because the size-age relationships will differ. On the basis of both statistical[77] and ecological[76] grounds, the use of the mean of the five largest thalli on a substrate can be firmly recommended.

There are certain situations where the use of a mean value should not be used. One of these is when the size-age relationship has been derived from gravestones. Gravestones represent particularly small surfaces and, frequently, there are only one or two thalli present on them.[76,78] Consequently, any mean value may be lower than expected and any derived size-age relationship should be based on the single largest thallus per gravestone.

C. Sample Area

The effects of sample area are much more difficult to assess. The optimum procedure appears to be the complete search of a surface, having first divided the surface into at least ten units. The data from the different units can then be compared and the presence of any anomalies in the data identified.

It may not always be possible to search entire deposits, and some form of sampling may have to be adopted. A number of procedures have been described in the literature. Beschel[65] recommended that an area of at least 100 m^2 be searched on each deposit. Larger areas have been used by Stork[43] and Mottershead,[79] who used 5 quadrants of 25 m^2 on each surface. Andrews and Webber[30] used 8 quadrants of 64 m^2 per surface, and Gordon[66] reported searching 600 m^2 per deposit. A "random walk" method has been used[10,33] whereby the largest lichens found during a fixed time period are recorded.

There has been virtually no work done on the reproducibility of such techniques, although Innes[77] has examined variations in different sizes of sample area. The basic problem is that if sample areas are too small, microenvironmental variations may affect the size of the thalli to a greater extent than age-related factors. As the sample size increases, the variations attributable to microenvironment will decrease. An opportunity for examining this is provided by the data collected on the Storbreen glacier foreland in southwest Norway by Matthews.[60,80,81] In these studies, the sample areas were 25 m by about 30 m and the entire moraine sequence was searched. The data from the outermost moraine have been examined statistically and, in Figure 3, the deviations from the overall mean are shown. The data from each side of the foreland were tested for serial correlation, but in both cases, this was found to be insignificant at $p = 0.01$ (west side, $K = 2.8084$; east side, $K = 2.0588$; Von Neumann ratio test of independence). These results suggest that microenvironmental factors do not extend beyond 25 m lengths along the moraine. Innes has obtained data from the same site, but in this case, the quadrats were 2×8 m, with the shorter length lying parallel to the moraine crest. A total of 100 adjacent quadrats were examined, and the deviations from the mean are shown in Figure 4. Significant serial correlation was present ($K = 1.4319$), suggesting that microenvironmental conditions affect quadrats of this size. The critical limit must lie somewhere between these two quadrat sizes but, to date, there have been no attempts to establish the position of the limit. It should be borne in mind that the reproducibility of quadrat measurements will be inversely related to the quadrat size due to the decrease in search efficiency in larger quadrats.

IV. ANOMALOUS THALLI

A problem that should be mentioned at this stage is the presence of anomalous thalli on a substrate. These may be either larger or smaller than expected from the age-size relationship. Coalesced thalli have already been discussed and will not be further examined. "Anomalously" large thalli may also occur due to environmental factors. Whether they should be regarded as anomalous is open to question. The aim of lichenometry is to measure the largest

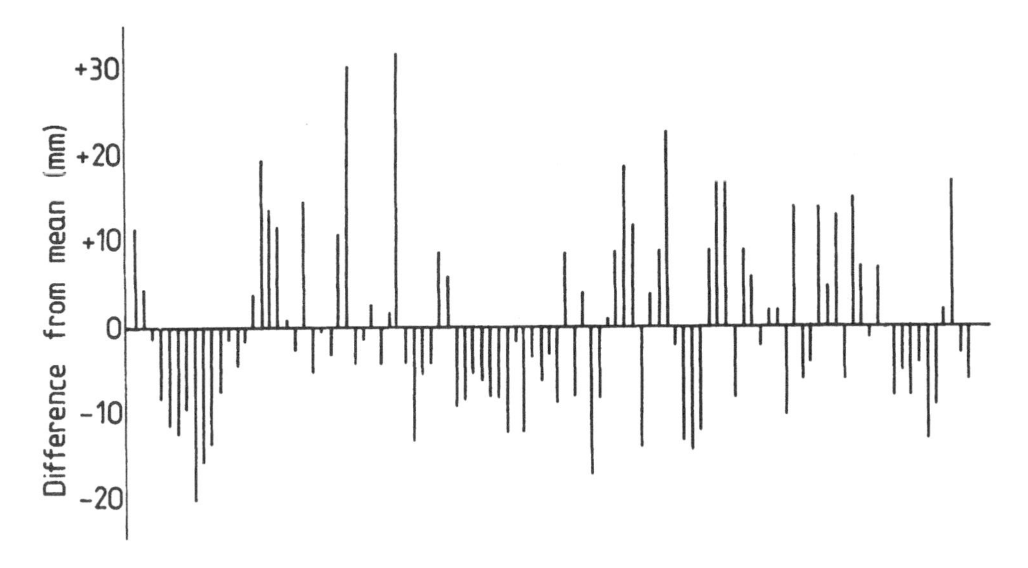

FIGURE 3. Differences between the mean of the five largest *Rhizocarpon* section *Rhizocarpon* thalli in individual quadrats and the mean value for the same index for all the quadrats at Storbreen, southwest Norway. Individual quadrat size = 750 m^2. All quadrats are from a moraine of uniform age (about 1750 A.D.).

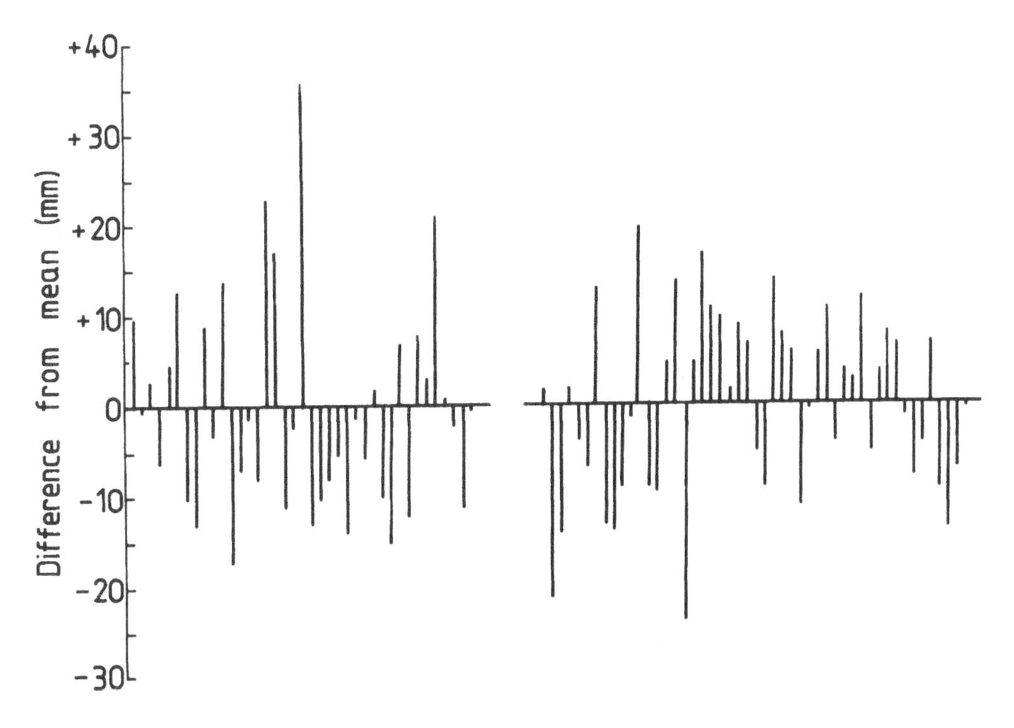

FIGURE 4. Differences between the mean of the five largest *Rhizocarpon* section *Rhizocarpon* thalli in 100 individual quadrats and the mean value for the same index for all the quadrats. Individual quadrat size = 16 m^2.

possible thalli on a substrate of a given age. The assumption is usually made that the environmental factors that result in the maximum age-size relationship are present on all surfaces. This may not be valid.

One of the most widely recognized problems is associated with the larger than average thalli that sometimes occur along stream margins.[53] This may be due to the enhancement of growth from the greater moisture availability or to the presence of faster-growing taxa on stream margins (e.g., *R. macrosporum* is known to favor stream margins and to grow faster than other *Rhizocarpon* section *Rhizocarpon* taxa.[48]). Innes[82] has examined this problem, but found little evidence of larger thalli on stream margins, particularly on young substrates. The study was conducted in southwest Norway and it is possible that the thalli are rarely moisture stressed for any length of time (the mean monthly precipitation is >60 mm). This would contrast with the many American and southern European alpine environments where significant moisture stress may occur during the growing season. The inclusion/exclusion of thalli growing along stream margins will therefore depend on (1) whether moisture stress is likely to be significant in the environment and (2) whether streams are present on all surfaces.

As certain environmental conditions may enhance growth, so they may depress it. Snow cover is of particular importance in alpine environments and may enhance growth by providing water well into the growing season. However, if it persists too long, it will inhibit photosynthesis and growth. The point at which enhancement changes to depression is not known. It should be remembered that snow cover was more extensive in the recent past and this may significantly affect the distribution of the largest lichens on a surface.

Other conditions may lead to anomalous thalli on a surface. For example, stone monuments may be regularly cleaned resulting in thalli that are smaller than expected. On natural surfaces, various mechanisms may result in the introduction of anomalous thalli. For example, avalanches may introduce "old" boulders onto a young surface and the lichens may survive the transition. If avalanche activity is particularly frequent, an old surface may be covered by fresh material and the date will be an underestimate of the age. Deposits may be reworked by fluvial activity or may be affected by instability. An important example of the latter is given by ice-cored moraines, which may be highly unstable for some time after they have been deposited. In such cases, the lichens will date only the onset of stability.

There is very little that can be done about "anomalous" thalli. It may be possible to identify and exclude particularly large ones, but if thalli are smaller than expected, only a minimum age can be obtained. In such cases, the value of any interpretation of the dates will depend on the experience of the observer. Most dating techniques (particularly ^{14}C assay) have a similar dependence on the interpretation of the derived date.

V. SIZE-AGE RELATIONSHIP

The use of the term "growth curve" have been avoided in this account. The majority of lichenometric studies are concerned with the establishment of curves around lichen data from surfaces of known age. Surfaces of unknown age are then dated by interpolation or extrapolation. Actual growth rates are not normally used at any stage. Each point on the curve represents the time that it has taken that thallus to reach the specified size. The nature of growth prior to this point is unknown. This means that factors such as colonization time and competition can be ignored, provided that they are uniform across the surfaces being investigated.

The above argument has important implications for the use of directly measured growth rates in lichenometry. Direct measurements are made over a specific time period and, if measurements of thalli of a range of sizes are taken, it may be possible to construct a growth curve.[83,84] It must be stressed that such curves are not directly applicable to lichenometric studies. This is because the curve refers to the growth of specified thalli during the measurement period only. The application of such a curve in lichenometry requires (1) that the length of the colonization period be known and (2) the assumption that thalli of a given size have grown at the same rate in the past. Such an assumption appears to be untenable in the light of much recent work on annual fluctuations in growth rates. Work in the Swiss Alps

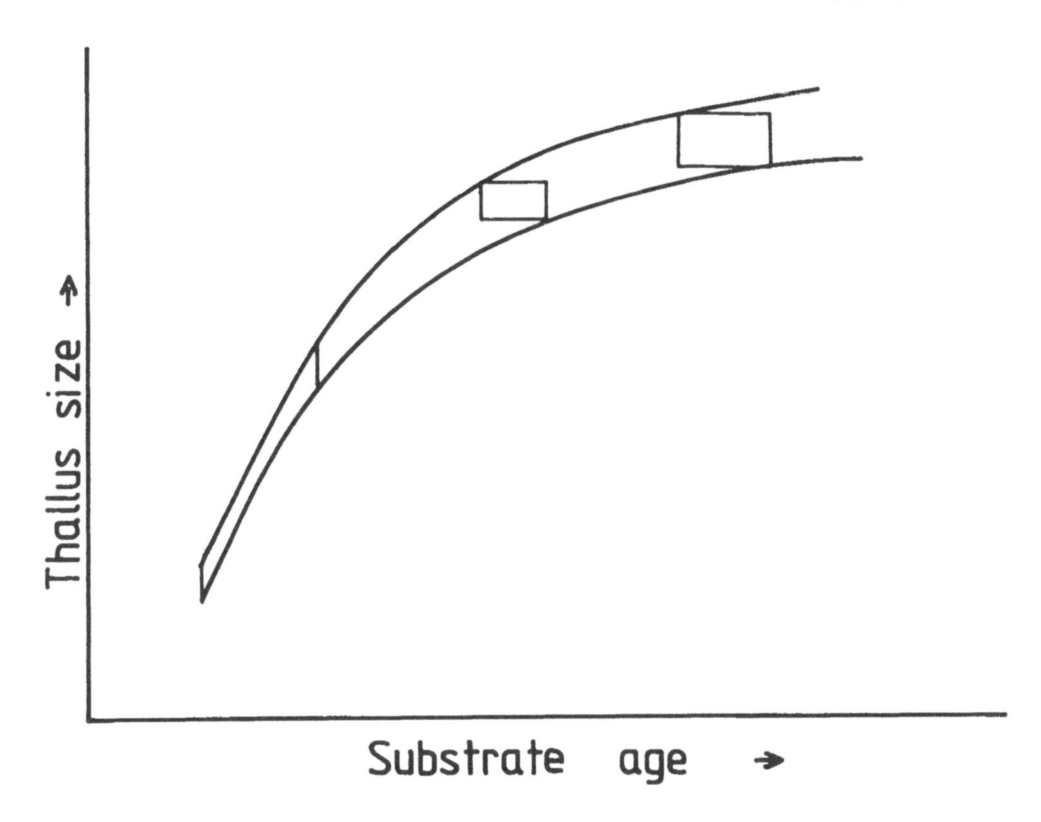

FIGURE 5. Development of an age-size relationship using ^{14}C dates. The two lower bars represent the error associated with the lichen measurements on surfaces of known age. The boxes are the error involved when surfaces are ^{14}C dated. The horizontal distance represents the error associated with the dating; the vertical distance is the error inherent in lichen measurements. Envelope curves should be used to portray the increasing potential for error as the age of the surface increases.

has suggested that directly measured growth rates can be used in lichenometry in some situations,[71] but much more work is needed before the method can be fully recommended.

In many studies, ''indirect'' methods have been used to determine the age-size relationship. These rely on locating the largest thallus (or thalli) on a surface and plotting it (them) as a function of surface age. The variation in the results will be directly related to the area of surface sampled. This can be illustrated by plots of data derived from graveyards.[20,28,41,78] The absence of a perfect linear relationship means that, normally, an envelope curve is drawn around the data. With increasing surface area, the scatter is reduced due to the decrease in microenvironmental variations between sites. Even so, age ''reversals'' may be discovered in lichenometric data when the relative age of surfaces is known.[41] Such reversals can usually be attributed to variations between the surfaces. In some cases, it may be possible to obtain dates using other lichenometric techniques (see below).

The ages of the control surfaces may be derived from a variety of sources. The most reliable are ''historical records'', in which the age of a surface is known from descriptions or photographs in the literature. In many studies, ^{14}C assay has been used to derive the date of deposition of a surface.[4,35] This technique is not recommended due to the potential errors involved. If ^{14}C dates are used, then they should be expressed in terms of ± 2 standard deviations, and should be calibrated using one of the high-precision calibration curves that are currently available. The errors should be shown on the final size-age relationship, with an envelope curve drawn around the limits (Figure 5).

An alternative method of constructing a size-age relationship should be mentioned here. This is based on the direct measurement of fast-growing taxa followed by the comparison of the diameters of fast- and slow-growing taxa on given deposits. This approach has been adopted in particular by Andrews and Webber,[30] Miller,[33] and Miller and Andrews.[34] For example, the *Rhizocarpon* curve for East Baffin Island is partly based on this approach.[34] The relative growth rates of *Pseudephebe minuscula* and "*Rhizocarpon geographicum*" were established for 140 sites of varying age. The growth rate of *P. minuscula* was established by direct measurements of 8 thalli over a 2-year period. Using these data, together with control data from a graveyard site and a ^{14}C-dated surface, a "*Rhizocarpon geographicum*" age-size relationship was developed. The technique has some potential, but the value of the above study is limited by the small sample of thalli used in the direct growth measurements. However, the technique combines the potential error of both direct and indirect measurements, and research is required to establish its validity.

The precise procedures used in the construction of age-size relationships vary and appear to be a matter of choice, which in turn will depend on the nature of the problem. However, the lack of standardization in the past has meant that the majority of studies are not directly comparable. The procedure that I favor follows:

1. Each surface should be divided into at least ten quadrats and the longest axes of the largest ten thalli of the target taxa recorded in each.
2. Anomalously large thalli are identified from this data set by either subjective means or statistical methods. Any such thalli are then excluded.
3. The mean of the largest five thalli on each surface is used to characterize the age of that surface.

This method does not enable confidence limits to be assessed. It would be possible to calculate the limits if the data from the different quadrats were not combined at stage 3 above. A mean value could be obtained for the mean of the five largest lichens from each quadrat. For such a value to be meaningful, the quadrats from every surface should be of equal size. The problem with such an approach is that the effects of microenvironmental variations will be introduced into the analysis. The variations will steadily become more apparent as the total area of each surface decreases. If the distribution of microenvironments is uneven, then quadrat means may be adversely affected.

Both methods have their advantages and disadvantages. At present, there is not enough information available to state which of the two methods is better. Further work is required in this field.

Finally, lichen thalli do not continue living forever. At some stage, senescence and mortality must occur. The point at which this happens will depend on (1) the growth rate of the lichen and (2) the nature of the substrate. At present, there is little evidence for any form of cyclic equilibrium, whereby older thalli die and are replaced by new colonies. In southwest Norway, surfaces of about 9000 years of age have little *Rhizocarpon* cover, suggesting that the dominance of *Rhizocarpon* thalli on some relatively young surfaces is only a temporary phase within a succession. This requires further investigation.

A. Shape of the Relationship

A variety of size-age relationships have been proposed, although most of these can be divided into either linear or negative exponential curves. Linear relationships have been proposed by Burrows and co-workers[85-87] in New Zealand, but the age control on the surfaces is now known to be incorrect.[73] Linear relationships have also been proposed for parts of Italy,[88] Iceland,[61,62] and Switzerland.[71] As these deal with fairly small thalli (<80 mm), it is possible that they represent the lower portion of a negative exponential curve. The negative exponential relationship is the same as has been used widely to represent growth curves in lichenology, and it is probably the most appropriate in the majority of situations. However,

Table 3
DIFFERENT TYPES OF FUNCTION THAT COULD BE USED TO DESCRIBE THE AGE-SIZE RELATIONSHIP

Linear	$y_t = a + bx$
Power	$y_t = ax^b$
Negative exponential	$\log (y_t + c) = a + bx$
Logistic	$y_t = 1/(a + bc^x)$
Gompertz	$y_t = ab^{c^x}$

Table 4
FACTORS THAT SHOULD BE CONSIDERED WHEN APPLYING A SIZE-AGE RELATIONSHIP DERIVED FOR ONE SURFACE TO ANOTHER

Substrate lithology
Moisture availability
Substrate stability
Snow cover
Vegetation cover
Altitude
Size of boulders
Temperature
Light intensity
Exposure to wind
Pollution
Aspect
Competition

a variety of models are available (Table 3), and the best procedure appears to be the adoption of whichever model fits the data best. The use of a mathematical model is recommended since it reduces the element of subjectivity involved in envelope curves drawn by eye. Furthermore, it may enable the extrapolation of size-age relationships beyond the period for which there is age control. However, extrapolations should only be made with extreme care.

A last point that should be made is that the appropriateness of the models listed in Table 3 is normally assessed by goodness-of-fit tests based on regression. This assumes a normal distribution of errors, which may not always be tenable. A potentially powerful technique of model fitting is provided by the use of generalized linear models, many of which do not require such an assumption. This technique is yet to be exploited.

B. Application of Size-Age Relationships

One of the greatest difficulties facing lichenometry is the application of size-age relationships from one area to another. This is often necessary when there is no dating control for any of the surfaces being investigated and surfaces from outside the immediate study area have to be used. The spatial transfer of the relationships rests on the assumption of similar environmental conditions between the sites. The factors influencing the relationship recently have been examined in detail by Innes[1] and are not restated here. The available evidence suggests that major variations in the relationship occur on both the macro- and microscale, but mesoscale variations appear to be relatively small. The factors that should be held as constant as possible are given in Table 4. If any of these vary between sites, an attempt should be made to assess their effect on the size-age relationships at the sites being investigated.

VI. OTHER METHODS

Two further techniques involved with the use of lichens in dating should be discussed. These are percentage-cover and size-frequency studies. Both are based on the premise that regular changes occur through time and that, if quantified, these changes can be used to date surfaces.

A. Percentage-Cover Measurements

The percentage cover of lichens has been used as a relative dating technique for some time.[10,27,58,59] Estimates are made visually, with Locke et al.[2] providing charts upon which comparisons can be made. The number of boulders used per surface varies from 50 to 250. Several indices can be calculated from such data, with the mean percentage cover on all boulders in the sample and the maximum percentage cover on any one boulder being the two most frequently used values.

As with the largest lichen diameters, the percentage cover is strongly influenced by environmental factors and care should be taken to ensure that surfaces are similar. This is particularly important when examining moraines, as the percentage cover may increase down the ridge in the same way as the largest diameters.[90-92]

The following procedures are recommended:

1. The total sample of boulders should not be less than 250 on each surface to be dated, each boulder having an upper surface of at least 500 cm^2 available for lichen colonization.
2. The sample should consist of several subsamples (>4) drawn from different locations on the surface (subject to step 5). The aim of this is to reduce the potential effect of an anomalous area.
3. On each boulder, the total lichen cover, the cover of the target taxa, and the cover of as many other taxa as possible should be recorded.
4. If the boulders on the surface consist of more than one lithology, sampling should be restricted to a single lithology. If more than one lithology is used, the lithology of each boulder should be noted and summary statistics calculated for each lithology.
5. Each subsample should be taken from the same position on the substrate. Thus, on moraines, the sampling position should be consistent (e.g., ridge crests or ridge bases).
6. Mean values should be used for each surface (e.g., mean *Rhizocarpon* section *Rhizocarpon* cover and mean total cover. Means should be given together with the 95% confidence limits. The maximum cover on any one boulder does not appear to be a particularly useful index, especially when absolute dates are sought.
7. All observations should be made by a single observer since they are highly subjective, and the results obtained by different observers are significantly different. The consistency of the observations improves with practice, and I recommend that a trial survey, using at least 500 boulders, be made before data are collected for the main study.
8. In some cases, the relative percentage cover (% cover of target taxa × 100/total lichen % cover on boulder) is a better index of age than the absolute percentage cover.[27] However, not enough is known about the properties of either to make any firm recommendation.

The reproducibility of the technique could be substantially improved by the use of objective measurements. Percentage cover could be measured using cross-wires or pins, but this would be extremly time consuming. Alternatively, it could be measured by the digital image analysis of photographs of the boulders. This would provide a reliable but rather expensive method of obtaining percentage-cover data.

The majority of studies have used percentage cover as a relative dating technique and few workers have attempted to use it for absolute dating. The technique has some potential for absolute dating, but more work is required on the methodology before it can be fully accepted.[91]

B. Size-Frequency Studies

Another method that has been used to date surfaces is the size-frequency characteristics of lichen population subsamples.[93,94] Locke et al.[2] recommended the use of the 1-in-1000

FIGURE 6. Storbreen, southern Norway. The moraine ridges appear as dark lines. This site has been the subject of numerous lichenometric studies.[1,55,56,60,77,80-82,91-93]

thallus to summarize these. This index is the thallus diameter that is predicted when a log-linear regression is run through a size-frequency distribution of 1000 thallus measurements. The model assumes that the size-frequency distribution is distributed in a log-linear fashion. This is not always the case, and the model is only appropriate in some situations. However, the technique has considerable potential, especially if other models fitting the distribution are developed. Again, the use of generalized linear models could be of great value.

The following procedures can be recommended:

1. Several samples (>9) should be taken from each substrate, each sample being from a similar position on the substrate.
2. The size of each sample will depend on the mean thallus size and the sampling interval. In most cases, subsamples of 1000 thalli should be sufficient.
3. Frequently, it may be practical to measure and record thalli to the nearest 1 mm. However, data recorded at this interval exhibit numerous irregularities which may obscure more important trends. Also, considerable observer bias may occur at this interval. Consequently, diameters should be reported to the nearest 5 mm.
4. Thalli of <10 mm diameter should be excluded, if possible. It may be necessary to include them if the maximum diameter is <40 mm, but they are often so abundant that they severely distort the frequency distributions.
5. The use of an index to characterize the distribution will depend on the nature of that distribution. The 1-in-1000 thallus, determined by log-linear regression, is only appropriate in some distributions. When this model is inappropriate, it may be possible to use an alternative model to calculate the 1-in-1000 thallus. Whether predicted values obtained from different models are directly comparable is unknown. Alternative indices include the mean, mode, median, standard deviation, and skewness. The relative values of these are unknown.
6. For each surface, the mean index value should be calculated and given together with 95% confidence limits.

Many lichenometric studies were carried out at Storbreen in southern Norway (Figure 6).

VII. CONCLUSIONS

The use of lichens in dating has come a long way since the pioneering work of Roland Beschel. The technique has been applied widely, although several studies have severely abused the technique. Lichenometry provides a *possible* means of obtaining a date for a surface. It is by no means universally applicable. Surfaces may be unsuitable for a number of reasons, including lithology, age and environmental conditions at the site. Many of the problems surrounding the technique have now been resolved, but several, detailed in the above account, remain. Provided that the technique is applied correctly and with care, dates accurate to about 10% of the surface age should be possible and, in some cases, it may be feasible to obtain more accurate dates. As such, the technique is of considerable value in a variety of situations where other dating techniques might be unsuitable or inapplicable.

REFERENCES

1. **Innes, J. L.,** Lichenometry, *Prog. Phys. Geogr.*, 9, 187, 1985.
2. **Locke, W. W., III, Andrews, J. T., and Webber, P. J.,** A manual for lichenometry, *Br. Geomorphol. Res. Group Tech. Bull.*, 26, 1, 1979.
3. **Curry, R. R.,** Holocene climatic and glacial history of the central Sierra Nevada, California, *Geol. Soc. Am. Spec. Pap.*, 123, 1, 1969.
4. **Karlén, W. and Denton, G. H.,** Holocene glacier fluctuations in Sarek National Park, northern Sweden, *Boreas*, 5, 25, 1975.
5. **Follman, G.,** Lichenometrische Alterbestimmungen an vorchristlichen Steinsetzungen der Polunesischen Osterinsel, *Naturwissenschaften*, 48, 627, 1961.
6. **Beschel, R. E.,** Lichenometrie im Gletschervorfeld, *Jahrb. Ver. Schutz Alpenpflazen -tiere München*, 22, 164, 1957.
7. **Winchester, V.,** A proposal for a new approach to lichenometry, *Br. Geomorphol. Res. Group Tech. Bull.*, 33, 1, 1984.
8. **Jochimsen, M.,** Does the size of lichen thalli really constitute a valid measure for dating glacial deposits?, *Arct. Alp. Res.*, 5, 417, 1973.
9. **Bornfeldt, F. and Österborg, M.,** Lavarter som hjälpmedel för datering av ändmoräner vid norska glaciärer, Department of Geography, University of Stockholm, Sweden, 1958.
10. **Birkeland, P. W.,** The use of relative age-dating methods in a stratigraphic study of rock glacier deposits, Mt. Sopris, Colorado, *Arct. Alp. Res.*, 5, 401, 1973.
11. **Beschel, R. E.,** Richerche lichenometriche sulle morene del Gruppo del Gran Paradiso, *Nuovo G. Bot. Ital.*, 65, 538, 1958.
12. **Beschel, R. E.,** Flechten als Altersmaβtab rezenter Moränen, *Z. Gletscherkd. Glazialgeol.*, 1, 152, 1950.
13. **Trudgill, S. T., Crabtree, R. W., and Walker, P. J. C.,** The age of exposure of limestone pavements — a pilot lichenometric study in Co. Clare, Eire, *Trans. Br. Cave Res. Assoc.*, 6, 10, 1979.
14. **Miller, C. D.,** Chronology of Neoglacial moraines in the Dome Peak area, North Cascade Range, Washington, *Arct. Alp. Res.*, 1, 49, 1969.
15. **Vareschi, V.,** Lichenologische Beiträge zur Eiszeitproblemen in den Anden, *Dtsh Bot. Ges. N. F.*, 4, 81, 1970.
16. **Orombelli, G. and Porter, S. C.,** Lichen growth curves for the southern flank of the Mont Blanc Massif, Western Italian Alps, *Arct. Alp. Res.*, 15, 193, 1983.
17. **Beschel, R. E. and Weidick, A.,** Geobotanical and geomorphological reconnaissance in West Greenland, 1961, *Arct. Alp. Res.*, 5, 311, 1973.
18. **Benedict, J. B.,** Recent glacial history of an alpine area in the Colorado Front Range, U.S.A. I. Establishing a lichen growth curve, *J. Glaciol.*, 6, 817, 1967.
19. **Benedict, J. B.,** Recent glacial history of an alpine area in the Colorado Front Range, U.S.A. II. Dating the glacial deposits, *J. Glaciol.*, 7, 77, 1968.
20. **Carrara, P. E. and Andrews, J. T.,** Problems and application of lichenometry to geomorphic studies, San Juan Mountains, Colorado, *Arct. Alp. Res.*, 5, 373, 1973.
21. **Miller, C. D.,** Chronology of Neoglacial deposits in the Northern Sawatch Range, Colorado, *Arct. Alp. Res.*, 5, 385, 1973.

22. **Carrara, P. E. and Andrews, J. T.**, Holocene glacial/periglacial record; northern San Juan Mountains, southwestern Colorado, *Z. Gletscherkd. Glazialgeol.*, 11, 155, 1976.
23. **Mahaney, W. C.**, Neoglacial chronology in the Fourth of July Cirque, Central Colorado Front Range, *Geol. Soc. Am. Bull.*, 84, 161, 1973.
24. **Løken, O. H. and Andrews, J. T.**, Glaciology and chronology of fluctuations of the ice margin at the south end of the Barnes Ice Cap, Baffin Island, NWT, *Geogr. Bull.*, 8, 341, 1966.
25. **Duford, J. M. and Osborn, G. D.**, Holocene and latest Pleistocene cirque glaciations in the Shuswap Highland, British Columbia, *Can. J. Earth Sci.*, 15, 865, 1978.
26. **Nikonov, A. A. and Shebalina, T. Y.**, Lichenometry and earthquake determination in central Asia, *Nature (London)*, 280, 675, 1979.
27. **Innes, J. L.**, Relative dating of Neoglacial moraine ridges in North Norway, *Z. Gletscherkd. Glazialgeol.*, 20, 53, 1984.
28. **Gregory, K. J.**, Lichens and the determination of river channel capacities, *Earth Surf. Processes*, 1, 273, 1976.
29. **Beschel, R. E.**, Observations on the time factor in interactions of permafrost and vegetation, *Proc. 1st Can. Conf. Permafrost*, Tech. Memorandum No. 76, National Research Council Canada, 1963, 43.
30. **Andrews, J. T. and Webber, P. J.**, A lichenometrical study of the northwestern margin of the Barnes Ice Cap: a geomorphological technique, *Geogr. Bull.*, 22, 80, 1964.
31. **Andrews, J. T. and Webber, P. J.**, Lichenometry to evaluate changes in glacial mass budgets as illustrated from north-central Baffin Island, *Arct. Alp. Res.*, 1, 181, 1969.
32. **Harrison, D. A.**, A reconnaissance glacier and geomorphological survey of the Duart Lake area, Bruce Mountains, Baffin Island, N.W.T., *Geogr. Bull.*, 22, 57, 1964.
33. **Miller, G. H.**, Variations in lichen growth from direct measurements: preliminary curves for *Alectoria minuscula* from eastern Baffin Island, N.W.T., Canada, *Arct. Alp. Res.*, 5, 33, 1973.
34. **Miller, G. H. and Andrews, J. T.**, Quaternary history of Northern Cumberland Peninsula, East Baffin Island, NWT, Canada. VI. Preliminary lichen growth curve, *Geol. Soc. Am. Bull.*, 83, 1133, 1972.
35. **Calkin, P. E. and Ellis, J. M.**, Development and application of a lichenometric dating curve, Brooks Range, Alaska, in *Quarternary Dating Methods*, Mahaney, W. C., Ed., Elsevier, Amsterdam, 1984, 227.
36. **Ellis, J. M., Hamilton, T. D., and Calkin, P. E.**, Holocene glaciation of the Arrigetch Peaks, Brooks Range, Alaska, *Arctic*, 34, 158, 1981.
37. **Harrison, D. A.**, Recent fluctuations of the snout of a glacier at McBeth Fiord, Baffin Island, NWT, *Geogr. Bull.*, 8, 48, 1966.
38. **Denton, G. H. and Karlén, W.**, Lichenometry: its application to Holocene moraine studies in southern Alaska and Swedish Lapland, *Arct. Alp. Res.*, 5, 347, 1973.
39. **Moiroud, A. and Gonnet, J.-F.**, *Jardins des Glaciers*, Editions Allier, Grenoble, 1977.
40. **Innes, J. L.**, Lichenometric dating of debris flow activity in the Scottish Highlands, *Earth Surf. Processes Landforms*, 8, 579, 1983.
41. **Innes, J. L.**, Lichenometric dating of moraine ridges in northern Norway: some problems of application, *Geogr. Ann.*, 66A, 341, 1984.
42. **Innes, J. L.**, Lichenometric dating of debris-flow deposits on alpine colluvial fans in southwest Norway, *Earth Surf. Processes Landforms*, 10, 519, 1985.
43. **Stork, A.**, Plant immigration in front of retreating glaciers, with examples from the Kebekajse area, northern Sweden, *Geogr. Ann.*, 45, 1, 1963.
44. **King, L. and Lehmann, R.**, Beobachtungen zur Oeklogie und Morphologie von *Rhizocarpon geographicum* (L.) DC. und *Rhizocarpon alpicola* (Hepp.) Rabenh. im Gletschervorfeld des Steingletschers, *Berichte Schweiz. Bot. Ges.*, 83, 139, 1973.
45. **Burrows, C. J.**, Studies on some glacial moraines in New Zealand. II. Ages of moraines of the Mueller, Hooker, and Tasman glaciers, *N.Z. J. Geol. Geophy.*, 16, 831, 1973.
46. **Burrows, C. J. and Lucas, J.**, Variations in two New Zealand glaciers during the past 800 years, *Nature (London)*, 216, 467, 1967.
47. **Luckmann, B. H.**, Lichenometric dating of Holocene moraines at Mount Edith Cavell, Jasper, Alberta, *Can. J. Earth Sci.*, 14, 1809, 1977.
48. **Osborn, G. D. and Taylor, J.**, Lichenometry on calcareous substrates in the Canadian Rockies, *Quat. Res.*, 5, 111, 1975.
49. **Haeberli, W., King, L., and Flotron, W.**, Surface movement and lichen-cover studies at the active rock glacier near the Grubengletscher, Wallis, Swiss Alps, *Arct. Alp. Res.*, 11, 421, 1979.
50. **Reger, R. D. and Péwé, T. L.**, Lichenometric dating in the Central Alaska Range, in *The Periglacial Environment*, Péwé, T. L., Ed., McGill-Queen's University Press, Montreal, 1969, 223.
51. **Larsson, L. and Logewall, B.**, Studier framför glaciärer i Kebnekajsefjällen, unpublished report. Department of Geography, University of Stockholm, Sweden, 1958.
52. **Lindsay, D. C.**, Estimates of lichen growth rates in the maritime Antarctic, *Arct. Alp. Res.*, 5, 341, 1973.

53. **Beschel, R. E.,** Botany: and some remarks on the history of vegetation and glacerization, in *Jacobsen-McGill Arctic Research Expedition to Axel-Heiberg Island, Preliminary Report 1959-60,* Müller, B. S., Ed., McGill University, Montreal, 1961, 179.
54. **Černohorský, Z.,** *Rhizocarpon* Ram. em. Th. Fr., *Bestimmungschlüssel europäischer Flechten. Ergänzungsheft 1,* Poelt, J. and Vězda, A., Eds., J. Cramer, Vaduz, 1977, 217.
55. **Innes, J. L.,** Lichenometric use of an aggregated *Rhizocarpon* "species", *Boreas,* 11, 53, 1983.
56. **Innes, J. L.,** Use of an aggregated *Rhizocarpon* "species" in lichenometry: an evaluation, *Boreas,* 12, 183, 1983.
57. **Andrews, J. T. and Barnett, D. M.,** Holocene (Neoglacial) moraine and Proglacial lake chronology, Barnes Ice Cap, N.W.T., Canada, *Boreas,* 8, 341, 1979.
58. **Carroll, T.,** Relative dating techniques — a late quarternary chronology, Arikaree Cirque, Colorado, *Geology,* 2, 321, 1974.
59. **Barnett, D. M.,** Development, landforms and chronology of Generator Lake, Baffin Island, N.W.T., *Geogr. Bull.,* 9, 169, 1967.
60. **Matthews, J. A.,** Families of lichenometric dating curves from the Storbreen gletschervorfeld, Jotunheimen, Norway, *Nor. Geogr. Tidsskr.,* 28, 215, 1974.
61. **Caseldine, C. J.,** Resurvey of the margins of Gljúfurájökull and the chronology of recent deglaciation, *Jokull,* 33, 111, 1983.
62. **Gordon, J. E. and Sharp, M. J.,** Lichenometry in dating recent glacial landforms and deposits, southeast Iceland, *Boreas,* 12, 191, 1983.
63. **Hole, N. and Sollid, J. L.,** Neoglaciation in western Norway — preliminary results, *Nor. Geog. Tidsskr.,* 33, 213, 1979.
64. **Gribbon, P. W. F.,** Glaciological notes from Sukkertoppen, west Greenland, *J. Glaciol.,* 6, 752, 1967.
65. **Beschel, R. E.,** Dating rock surfaces by lichen growth and its application to glaciology and physiography (lichenometry), in *Geology of the Arctic,* Vol. 2, Raasch, G. O., Ed., University of Toronto Press, Toronto, 1961, 1044.
66. **Gordon, J. E.,** Glacier margin fluctuations during the 19th and 20th centuries in the Ikamiut Kangerdluarssuat area, West Greenland, *Arct. Alp. Res.,* 13, 47, 1981.
67. **Calkin, P. E. and Ellis, J. M.,** A Neoglacial chronology for the Central Brooks Range, Alaska, *Geol. Soc. Am. Abstr. Prog.,* 10, 376, 1978.
68. **Innes, J. L.,** A standard *Rhizocarpon* nomenclature for lichenometry, *Boreas,* 14, 83, 1985.
69. **Griffey, N. J.,** A lichenometrical study of the Neoglacial end moraines of the Okstindan glaciers and comparisons with similar recent studies, *Nor. Geogr. Tidsskr.,* 31, 163, 1977.
70. **Gellatly, A. F.,** Revised dates for two recent moraines of the Mueller Glacier, Mt. Cook National Park (Note), *N.Z. J. Geol. Geophys.,* 26, 311—315, 1983.
71. **Proctor, M. C. F.,** Sizes and growth rates of thalli of the lichen *Rhizocarpon geographicum* on the moraines of the Glacier de Valsorey, Valais, Switzerland, *Lichenologist,* 15, 249, 1983.
72. **Griffey, N. J.,** Investigation of the Neoglacial deposits of the Okstindan glaciers, in *Okstindan Research Project 1973 Preliminary Report,* Parry, R. B. and Worsley, P., Eds., University of Reading, Reading, Mass., 1975, 1.
73. **Birkeland, P. W.,** Soil data and the shape of the lichen growth-rate curve for the Mt. Cook area (Note), *N.Z. J. Geol. Geophys.,* 24, 443, 1981.
74. **Webber, P. J. and Andrews, J. T.,** Lichenometry: a commentary, *Arct. Alp. Res.,* 5, 295, 1973.
75. **Calkin, P. E. and Ellis, J. M.,** A lichenometric dating curve and its application to Holocene glacier studies in the Central Brooks Range, Alaska, *Arct. Alp. Res.,* 12, 245, 1980.
76. **Innes, J. L.,** An examination of some factors affecting the largest lichens on a substrate, *Arct. Alp. Res.,* 17, 98, 1985.
77. **Innes, J. L.,** The optimal sample size in lichenometric studies, *Arct. Alp. Res.,* 16, 233, 1984.
78. **Innes, J. L.,** The development of lichenometrical dating curves for Highland Scotland, *Trans. R. Soc. Edinburgh Earth Sci.,* 74, 23, 1983.
79. **Mottershead, D. N.,** Lichenometry — some recent applications, in *Timescales in Geomorphology,* Cullingford, R. A., Davidson, D. A., and Lewin, J., Eds., John Wiley & Sons, New York, 1980, 95.
80. **Matthews, J. A.,** Experiments on the reproducibility and reliability of lichenometric dates, Storbreen gletschervorfeld, Jotunheimen, Norway, *Nor. Geogr. Tidsskr.,* 29, 97, 1975.
81. **Matthews, J. A.,** A lichenometric test of the 1750 end-moraine hypothesis: Storbreen gletschervorfeld, southern Norway, *Nor. Geogr. Tidsskr.,* 31, 129, 1977.
82. **Innes, J. L.,** Moisture availability and lichen growth: the effects of snow cover and streams on lichenometric measurements, *Arct. Alp. Res.,* in press.
83. **Hill, D. J.,** The growth of lichens with special reference to the modelling of circular thalli, *Lichenologist,* 13, 265, 1981.

84. **Schroeder-Lanz, H.,** Establishing lichen growth curves by repeated size (diameter) measurements in lichen individua in a test area — a mathematical approach, in *Late- and Postglacial Oscillations of Glaciers: Glacial and Periglacial Forms,* Schroeder-Lanz, H., Ed., A. A. Balkema, Rotterdam, 1983, 393.
85. **Burrows, C. J.,** Late Pleistocene and Holocene moraines of the Cameron Valley, Arrowsmith Range, Canterbury, New Zealand, *Arct. Alp. Res.,* 7, 125, 1975.
86. **Burrows, C. J. and Orwin, J.,** Studies on some glacial moraines in New Zealand. I. The establishment of lichen-growth curves in the Mount Cook area, *N.Z. J. Sci.,* 14, 327, 1971.
87. **Burrows, C. J. and Maunder, B. R.,** The recent moraines of the Lyell and Ramsay Glaciers, Rakaia Valley, Canterbury, *R. Soc. N.Z. J.,* 5, 479, 1975.
88. **Belloni, S.,** Richerche lichenometriche in Valfurve e nella Valle di Solda, *Boll. Com. Glaciol. Ital.,* 21, 19, 1973.
89. **Carrara, P. E. and Andrews, J. T.,** The Quarternary history of northern Cumberland Peninsula, Baffin Island, N.W.T. I. The late- and Neoglacial deposits of the Akudlermuit and Boas Glaciers, *Can. J. Earth Sci.,* 9, 403, 1972.
90. **Innes, J. L.,** The use of percentage-cover measurements in lichenometric dating, *Arct. Alp. Res.,* 18, 209, 1986.
91. **Innes, J. L.,** Zonation of lichens on Neoglacial moraine ridges at Storbreen, Jotunheimen, southwest Norway, unpublished manuscript.
92. **Haines-Young, R. H.,** Size variation of *Rhizocarpon* on moraine slopes in southern Norway, *Arct. Alp. Res.,* 15, 295, 1983.
93. **Innes, J. L.,** Size frequency distributions as a lichenometrical technique: an assessment, *Arct. Alp. Res.,* 15, 285, 1983.
94. **Andersen, J. L. and Sollid, J. L.,** Glacial chronology and glacial geomorphology in the marginal zones of the glaciers, Midtdalsbreen and Nigardsbreen, south Norway, *Nor. Geogr. Tidsskr.,* 25, 1, 1971.

Chapter XII.B

MEDICINAL AND OTHER ECONOMIC ASPECTS OF LICHENS

David H. S. Richardson

I. INTRODUCTION

Lichens have been used in medicine from the time of the early Chinese and Egyptian civilizations until the present.[1,2] Man has also exploited lichens for various other purposes and occasionally suffered from allergies or gastric irritation as a result of their use.[3-5] This chapter is concerned with such aspects and with current use. Undoubtedly, the greatest amounts of lichen are now used commercially in the perfume industry, amounting to some 9000 tons annually.[6] A remarkably large quantity — 2000 to 3000 tons — is collected every year by the Association of Finnish Lichen Exporters. The species involved is *Cladonia stellaris*, which is used mainly in Germany for the construction of Christmas and graveyard wreaths (Figure 1). However, it is also employed in model building and for decorations.[7]

II. ALLERGY

Dermatitis and eczema are used to describe a skin complaint characterized by reddening, scaling, and itching, which may be accompanied by swelling of the affected area. There is no general agreement as to the distinction between the two.[8] Contact dermatitis is an allergic response resulting from exposure to various chemicals. It is a common complaint and, if questioning does not elicit the probable cause, sufferers are frequently sent to departments of dermatology for patch testing against a standard group of allergens. Between 0.5 and 1% of those patients referred for tests show positive reactions (Figure 2) to lichen substances.[9-11] In general, those developing contact dermatitis from lichens show symptoms first on the hands or neck where the skin is exposed. It can spread to the arms (Figure 3) and face or even to the waistline or genitals where the allergens accumulate.[12]

Contact dermatitis among forestry and horticultural workers has been known for 65 years and is well documented, forming part of a syndrome known as "woodcutter's eczema" or "cedar poisoning".[5] Offending allergens are water and fat soluble and very resistant so that they can survive for years in dried form as, e.g., lichen-covered logs stored undercover for firewood. Among one group of sufferers studies, four were railway or road workers, two farmers, two were keen on outdoor pursuits, one was a carpenter, one was a teacher of botany, and one was a woman married to a forester. When patch tested, six of the group showed a stronger reaction when exposure to lichen substances was combined with light, indicating photocontact dermatitis. Among the many lichen substances responsible are usnic acid, evernic acid, fumaroprotocetraric acid, stictic acid, and atranorin.[13] Atranorin and stictic acid are capable of photosensitizing human skin as well as being contact allergens. Atranorin and other depsides of the β-orcinol type strongly absorb ultraviolet (UV) light and this may account for their ability to induce photocontact dermatitis. It is interesting that the forester's wife referred to above showed periodic periods of remission in winter when her husband did office work.[14,15]

Lichens are adversely affected by air pollution, particularly by sulfur dioxide, so that there are few species in city centers and the chance of urban dwellers developing lichen-induced dermatitis is much less. However, a number of cases of dermatitis have been traced recently to lichen substances in perfumes and aftershave lotions where lichen extracts are used as an important fixative for the fragrances (*vide infra*). If the lichen is extracted with

FIGURE 1. *Cladonia stellaris* used as a graveyard wreath in Germany. (Photo courtesy of M. R. D. Seaward.)

ethanol or subsequently treated with hot alcohol during extraction and purification, some of the allergens may be hydrolyzed to less potent forms.

The increasing use of perfumes and lotions in modern urban society suggests that this type of dermatitis will become more common. The clinical picture depends on the concentration of the allergen in the cosmetic. Pronounced reddening and swelling may occur in some cases at the site of application while in others only itching and a slight reddening are observed.[11] Three of four men sensitive to lichen substances developed facial dermatitis from aftershave lotions, while a nurse had dermatitis on her hands from applying the lotion to disabled patients. Another woman observed itching after dancing cheek to cheek and subsequently treated her face prophylactically with corticoid cream after dancing. Once sensitized to lichen-containing perfumes, dermatitis may follow exposure to lichens them-

FIGURE 2. A positive patch test in the form of a swollen red area following exposure of the skin to a perfume containing oakmoss (*Evernia prunastri*) extract. (Photo courtesy of S. Fregert.)

FIGURE 3. The arm of a woman patient showing the dermatitis that developed after using *Cladonia* (the lichen lying next to her arm) in a Christmas decoration. (Photo courtesy of S. Fregert.)

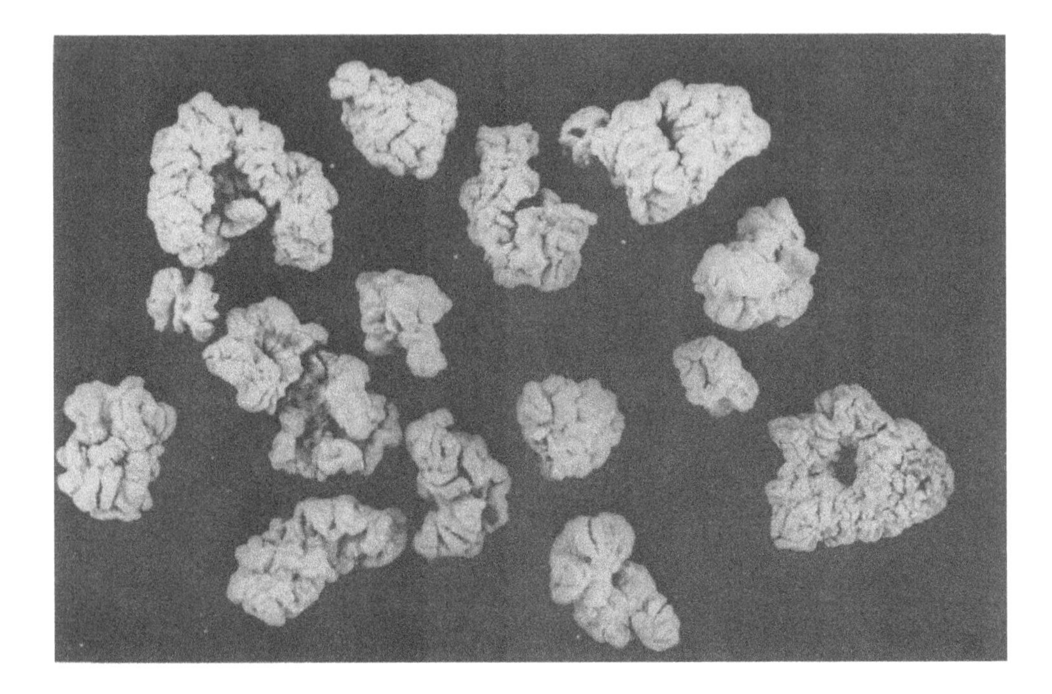

FIGURE 4. Detached fragments of *Lecanora esculenta* from Arabia which can be blown around and give this lichen the name "manna lichen".

selves. For example, handling lichen-covered branches, sitting on wet lichen-covered boulders, or arranging Christmas decorations incorporating *C. stellaris* (Figure 3) have led to relapses in patients who had given up using perfumes or lotions.[9,16]

III. FOOD

In the past, lichens were used very widely as food in times of hardship.[2,5,17,18] For example, in Norway during the years of bad harvest from 1807 to 1814, dried Iceland moss (*Cetraria islandica*) was used to supplement flour. It was considered a better substitute than bark flour (phloem tissue from elm, pine, or birch), which was also used to a great extent. The lichen was soaked in lye (aqueous extract of fresh wood ash) for 24 hr, which presumably neutralized the lichen acids. It was then dried and blended with grain before being ground into flour. Unfermented flat breads or porridge were usually made from the flour.[70]

In the semiarid lands of the Irano-Turanian steppes, highlands of North Africa, and deserts of west-central Asia are found *Lecanora* (section *Sphaerothallia*) *esculenta* and related species. These lichens form thick wrinkled crusts on rocks (Figure 4) which, as they become older, tend to become detached. They may be blown around by strong winds or washed into depressions by any accompanying rain showers. The sudden appearance of quantities of lichen from this cause has led to it being termed "manna lichen". Falls of manna lichen have been recorded by various observers (Figure 5). These large and easily exploited accumulations have been used as food by man or for his animals, usually sheep. Apparently during the French campaign in Algeria, *L. esculenta* was admixed with barley and fed to horses for 3 weeks without ill effect.[19,20] The lichen is reputed to contain a high proportion (up to 60% dry weight) of oxalic acid which might be expected to be a powerful irritant.

North American Indians used *Umbilicaria mühlenbergii* as a standby to make a soup when faced with starvation, while *U. esculenta* (Figure 6) is regarded as a delicacy in Japan, particularly in the area around Hiroshima. The latter is collected from very steep granite

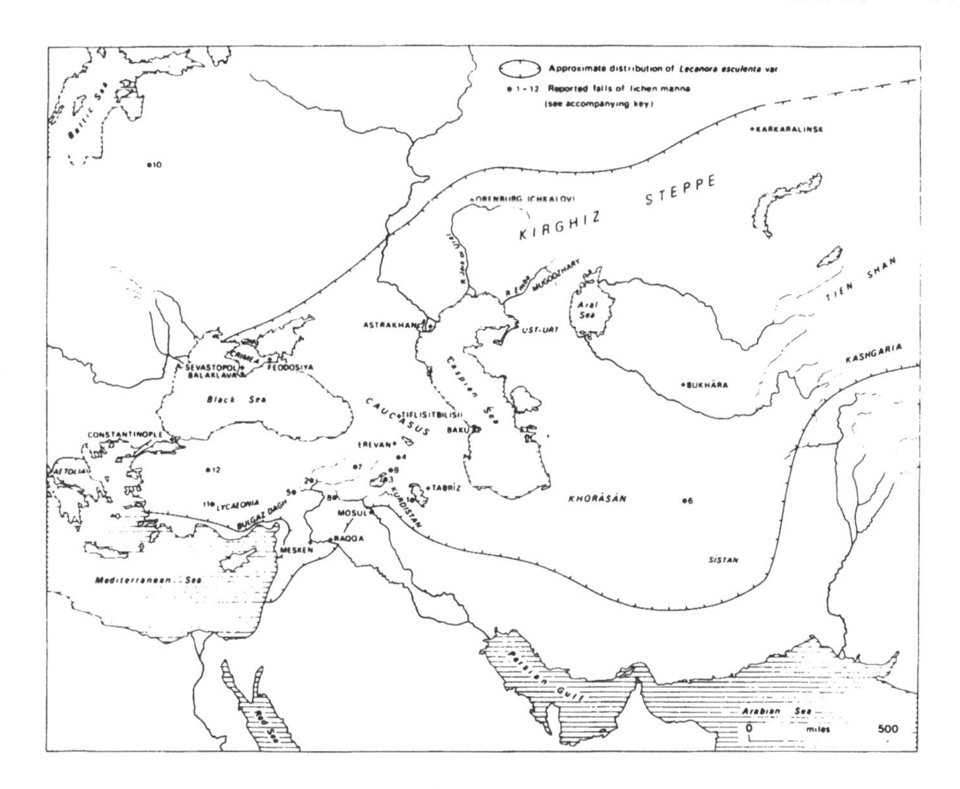

FIGURE 5. Falls of "manna lichen" at different times in western and central Asia involving accumulations of 10 to 15 cm of lichen. (1) 1829 and 1846; (2) 1864; (3) 1841 and 1864; (4) 1828; (5) 1864; (6) 1847; (7) 1857; (8) 1864 and 1890; (9) 1849; (10) 1846; (11) 1846; and (12) 1841. (From Donkin, R. A., *Anthropos*, 76, 562, 1981. With permission.)

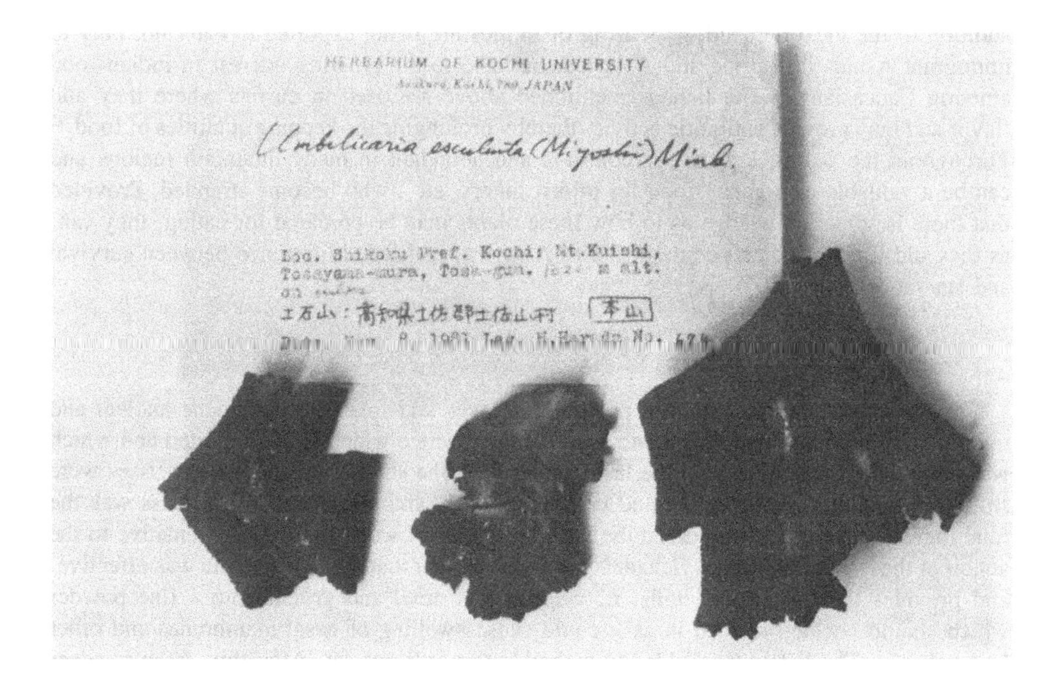

FIGURE 6. *Umbilicaria esculenta* ("Iwatake") eaten as a delicacy in the Hiroshima district of Japan. (Photo courtesy of M. R. D. Seaward.)

cliffs in the surrounding mountainous areas using climbing ropes; about 800 kg is eaten annually.[5,17] Another example of lichens as food is documented by Turner.[21,22] The interior peoples of western North America from British Columbia to northern California formerly used *Bryoria fremontii* as an important item in their diet. It is interesting that the Indians differentiate, by taste, the bitter vulpinic acid-containing *B. tortuosa* from *B. fremontii*, which is normally free of this lichen acid. This distinction was not generally recognized by lichen taxonomists until recently.[23] The nutritive value of *Bryoria* is higher in protein than *Cladonia* and many other lichens, and consists of 5% protein, 3% fat, 4% minerals, and a large crude fiber fraction that includes sugars, starches, pectins, and lignins that together total 40%.[24] After picking from the trees, the lichen is soaked in stream water and cooked in a steam pit. Hot rocks are placed in the bottom and covered with leaves and the mass of lichen. Soil is packed on top and removable stakes provide holes through which more water can be added. After cooking overnight or longer, the lichen is reduced to about a quarter of its original volume and can be removed and cut into loaves about 5 cm thick. These can be eaten at once or dried and stored for several years. Often layers of native onions (*Allium cernuum* or *A. geyeri*) or camas bulbs (*Camassia gwamash*) are added for flavor. Alternatively, berries and more recently sugar or molasses can be used as sweetners.[21]

Although lichens are no longer used as a major food item in man's diet, they are locally employed in herbal preparations (Figure 7) or to add flavor to food. Thus, *Parmelia austrosinensis, P. tinctorum,* and occasionally *Cetrariastrum nepalense* can be found on sale in Saudi Arabia and even in North London being imported from Nepal and Northern India.[71] Shops and street sellers in the markets of Poona and Aurangabad frequently stock "dagaful" (stone flowers). This is a mixture of *P. tinctorum, P. nilgherrense, P. reticulata,* and *P. sancti-algelia. Ramalina* and *Usnea* sometimes may be included. Garam Masala is a spice mixture which is usually added at the end of cooking to enhance the flavor. Kubal Garam Masala is a prepacked version, sold in India, in which one of the principal components is dagaful. The amount of material collected for the above products is probably large and must place a heavy burden on the diminishing macrolichen flora of the Indian subcontinent. In addition to the local demand, increasing quantities are being exported to cater not only to immigrant Asians in Europe and America, but also to the growing interest in Indian food amoung Caucasians.[72] The lichens mentioned above are used in curries where they add flavor and may provide antibiotic activity thereby prolonging the keeping qualtities of food.[25] Throughout the world, lichens are common and abundant in many mountain regions and can be a valuable emergency food for pilots, hikers, etc., who become stranded. Provided that there is some knowledge as to how these plants may be prepared for eating, they can, as they did for the Indian people of North America, make the difference between survival and starvation.

IV. POISONING

The use of *Letharia vulpina* for poisoning wolves and foxes appears to be ancient and traditional.[26,27] Smith[2] provides an account of this practice which is often quoted and which was based on two records from the last century.[28,29] She states that reindeer carcasses were stuffed with a mixture of lichen and powdered glass, and suggests that the glass was the fatal ingredient or at least rendered the internal organs of wolves extremely sensitive to the action of the lichen. However, Schade[30] has made it clear that the lichen alone was effective, and provides the following details. *L. vulpina* was dried and ground into a fine powder which should not be breathed in as it could cause swelling of nasal membranes and other hemorrhages. The lichen was added to melted butter or bacon fat. After this, fresh reindeer blood and meat was added and the whole mass allowed to solidify. The mixture was put in a portion of reindeer carcass between the skin and muscle. Wolves that ate the bait died within 24 hr.

FIGURE 7. An active tonic made from bark lichen, and commercially available in Europe. (Photo courtesy of M. R. D. Seaward.)

The toxic principle in *Letharia* is vulpinic acid and it poisons all meat eaters, but is not effective against mice and rabbits.[30] In cats, acute poisoning results in breathing difficulties, high blood pressure, and vomiting, whereas chronic effects induce internal bleeding and kidney damage. Cats are claimed to be more sensitive than hedgehogs, and even frogs are poisoned by as little as 4 mg of vulpinic acid.[2,30] The LD_{50} for a squirrel has been estimated as about 1.5 to 2.5 g of dry lichen, and several studies have shown that it is also toxic to insects and molluscs.[31,32] The thallus tips of *L. vulpina* contain the greatest concentration of vulpinic acid (about 4 to 6% dry weight), whereas the older basal branches have less than 2%. The former are metabolically most active and nutrient rich so that their high acid content would appear to provide a useful mechanism to deter herbivores. Field observations which indicate that insect damage to thalli of this lichen is rare provide circumstantial evidence for the operation of such a mechanism.[32,33] While the presence of high levels of

Table 1
THE QUANTITIES OF LICHENS PROCESSED ANNUALLY IN THE PERFUME INDUSTRY (MOXHAM, 1986)

Countries of collection	Tons
Yugoslavia	4,900—5,200
France	2,000—3,000
Morocco	900—1,000
India	120
Nepal	800
Unknown, but processed in India	200
Total	8,920—10,320

Note: No figures are available for Eastern Europe, Russia, or the United States.

From Moxham, T. H., *Progress in Essential Oil Research*, Brunke, E. J., Ed., Walter de Gruyter, Berlin, 1986. With permission.

lichen substances may discourage many insect and mollusc herbivores, recent studies have shown that certain lichen grazers can consume even food containing vulpinic acid without harm.[34] In North America, there seems to be no record of using *L. vulpina* in poisoned bait, but it was employed as a dye for porcupine quills incorporated into woven baskets and for blankets.[18,35]

V. PERFUMES

Preparations from *Evernia prunastri* (oakmoss) and *Pseudevernia furfuracea* (treemoss) are not only excellent odor fixatives, but also impart to the finished product — whether soap, cosmetics, lotion, or perfume — a peculiar earthy, mossy character. This provides the "bass note" of a perfume (just like a musical chord) and the "top notes" are the main flower ingredients and other constituents which combine to make up the whole fragrance. The alluring and exotic nature of a number of well-known and popular perfumes is owed to a substantial content of lichen extract.[36] This ability to act as both fixative and perfume has been known for centuries.[2,5]

The raw material required to make the various extracts is considerable (Table 1) and represents a vast bulk of collected lichens as these plants weigh very little when dry. Collection provides employment for large numbers of people in different countries, especially in the Massif Central of France, mountainous districts of Yugoslavia, Morocco, and North India.[37,38] Unfortunately, the financial return for the lichen pickers is very low in comparison with the high prices paid for the finished products.[37,39] The lichen is picked by hand from the lower branches, but scrapers are used for material above head height. In Yugoslavia, the canvas bags attached to the scrapers used there are periodically emptied into large jute sacks (Figure 8). These are transported to the packing station where the lichen is compressed into bales for shipping to the perfumeries. Incidentally, both the dust generated at this stage and the subsequent use of benzene in extraction pose potential occupational health problems.[37] A bale of commercial oakmoss is mainly *E. prunastri*, but several other species of fruticose lichens and a small quantity of foliose lichens are usually included. In contrast, treemoss samples (*P. furfuracea*) contain between 35 to 55% wood by weight and the different

FIGURE 8. Lichen scrapers with their collecting canvas bags used in Yugoslavia. Note the large lichen-filled hessian sack in the background used for transporting the lichen to the packing plant. (Photo courtesy of T. H. Moxham.)

fragrance from oakmoss is probably attributable to fractions derived from the tree bark.[6,37]

The extraction process used by the perfumeries is complex and the details kept secret.[5] Of the various solvents used, alcohol gives a high-quality extract, but has the disadvantage of low yield and costly working procedures. It has largely been replaced by petroleum ether and thiophene-free benzene. The resultant product (Figure 9) is a "concrete oil" which may be further refined by heating with warm alcohol that is then cooled to about 20°C for a period to yield the "absolute." The latter is free of waxes and other impurities which are either not soluble or precipitate out and are filtered off during the process. Extraction yields of the concrete vary depending on the solvent used but are usually between 2 to 5%, while the absolute derived from it amounts to 35 to 80%. Most companies offer a variety of oak- and treemoss extracts to satisfy the various odor requirements of perfumers.[40] Between 1 and 5% of the appropriate lichen extract is added to perfumes and other products.[41]

The use of lichens in the perfume industry is now the largest commercial exploitation of these plants and shows no signs of diminishing. Some concern is now expressed regarding sustained supplies as there is little systematic management of the lichen woodlands which are reharvested every 1 to 5 years. In addition, rural air pollution levels are rising as a result of the high stack emission policy adopted by most European industrial countries. The latter may result in reduced growth rates in these rather sensitive lichens and it eventually may become uneconomic to harvest from traditional collection areas.[37,39]

FIGURE 9. Collecting the lichen "concrete oil" after extraction of oakmoss, *Evernia prunastri*, for use in perfumes and soaps. (Photo courtesy of E. Zieghenbein, *Dragoco*.)

VI. DYES

Lichen dyes were only used widely for the commercial production of brown- and purple-colored woolen yarn. They continued to be used by professional dyers and weavers in the outlying areas of Scotland, Ireland, and Scandinavia long after being abandoned by major manufacturers in favor of synthetic dyes. Indeed, there are a few crofters who still produce Harris tweed that incorporates a proportion of yarn dyed with vegetable dyes including crottle from *Parmelia omphalodes*. This dye is extracted by a boiling water process and eventually the machine-carded wool is handwoven.[5] Today, it is amateur weavers who make most use of lichens, finding them a fascinating and varied source of dyes. However, many lichenologists have expressed concern at the use of species which have become rare as a result of rising rural air pollution combined with habitat destruction following felling or coniferous underplanting of deciduous woodland. This is because dyers use approximately 1 kg of dried lichen per kilo of wool so that a large amount of lichen is required even for a single woolen sweater. A number of practical guides have been published for the preparation of lichen dyes.[42-46] The yellow and brown dyes are extracted from lichens with boiling water. The dye is due to a reaction between free amino acids on wool or silk and aldehyde groups on the lichen substances forming stable azomethine linkages. The substances which give rise to brown and yellow dyes are very abundant, but often colorless until they break down on extraction. Thus, there is usually no correlation between the color of a lichen and the color which it yields on boiling with wool. Such dyes are color fast, unfading, and give the yarn a delightful aroma. Furthermore, dyed products are reputed to be mothproof as the lichen acids render the cloth distasteful to the caterpillars.[5,18]

The technique for producing red and purple dyes from lichens by steeping them in aged urine is very ancient.[26,47] The manufacture of this dye known as orchil or archel was later improved by the use of ammonia distilled from fermented urine. The raw material was a species of *Roccella*, called rock moss or weed, imported from the Canary Islands and elsewhere. In Britain, Cuthbert Gordon developed a process, in 1758, for extracting a similar dye from the native lichen *Ochrolechia tartarea* on a commercial scale. In an appeal to Parliament for financial support, Gordon noted that 504 tons of archel had been imported through the Port of London alone in the years 1781 to 1783. This was valued at approximately 100,000 pounds, whereas his native product "cudbear" (a contraction of his Christian name) would have cost less than 7,000 pounds. Thus, if his work were supported and dyers bought the local product, Great Britain would be saved a considerable sum.[48,49] He was evidently quite an entrepreneur as judged by his salesmanship recorded in letters whereby, on at least one occasion, he sent twice the ordered amount.[50] It is now known that purple dyes can be extracted from a range of lichens that contain such substances as gyrophoric acid, evernic acid, lecanoric acid, and erythrin. These are depsides which decompose to produce orcin. In the presence of ammonia and oxygen, this is converted to orecein, which is a mixture of 7-oxyphenoxazon, 7-aminophenoxazon, and 7-aminophenoxazin.[18]

A change in fashion, and later the invention of synthetic dyes, led to a decline in orchil manufacture by the middle of the last century.[5] A contributing factor may have been a tendency for orchil-dyed products to fade in bright sunlight, although the brilliance of the freshly dyed cloth could not be matched by synthetic dyes. The firm of Marshall & Harman, now known as Yorkshire Chemicals, produced cudbear and orchil for well over 100 years until 1940 when declining demand made manufacture impracticable.[51] As late as the 1950s, about 5 tons/year of lichen was still employed to produce orchil/cudbear by Messrs. Johnsons of Hendon mainly for export to America and about 5 times this amount for the production of litmus and orcinol.[5] However, the production of all these has now ceased in Britain, although litmus continues to be made at Wormer in the Netherlands where between 700 and 1400 kg of *Roccella* is used annually to produce 1 to 2 tons of litmus.[52] Litmus was produced in England by steeping *R. montagnei* from Madagascar, in large open tubs, in sodium carbonate solution and then adding ammonia. The mass was periodically stirred for about 1 month. The unwanted dyes were removed by extraction with alcohol in which they are soluble. The remaining fermented lichen was dried and ground up to form litmus powder.[5] The method used in the Netherlands is different and involves grinding up the lichen to a powder before adding the chemicals, but the source of the lichen and manufacturing details are secret.[52] Besides its use in schools as an acid-base indicator, litmus is also employed as a constituent of Henna hair conditioners manufactured by Milady® Tel-Aviv, Israel. These vegetable hair conditioners and tinting agents also contain extracts of henna, indigo, rhubarb, and Centaurea.

VII. MEDICINES

The use of lichens for medical purposes is of great antiquity and has been reviewed previously.[2,5,53] Various species were reputed to be effective in the treatment of coughs, ulcers, hemorrhages, thrush, worms, and rabies.[2,26] Even today, *Cetraria islandica* (Iceland moss) can be purchased in European pharmacies for the preparation of herbal medicines (Figure 10). However, handling moldy lichens can be hazardous as pulmonary fibrosis and other symptoms can ensue.[54]

In the 1960s, partly as a result of these traditional uses of lichens, considerable interest was shown by scientists in the chemical constituents of these plants. Although many lichen substances proved to be antibiotics, their activity was generally low compared with products from other microorganisms. In addition, lichen antibiotics were found to be too insoluble in water for therapeutic use. The most determined attempt to produce a commercially useful product came from Finland where "usno" was developed. This was a water-soluble com-

FIGURE 10. A series of products made from Iceland moss (*Cetraria islandica*) and beard lichen (*Usnea*). These were found on sale in pharmacies and health food shops in Switzerland, Germany, and Holland. Note the fragments of dried Iceland moss in the center of the photograph which frequently can be purchased loose from pharmacies throughout continental Europe. (Photo courtesy of D. H. S. Richardson.)

pound of usnic acid with benzyldimethyl-{2-[-(p-1,1,3,3-tetramethylbutylphenoxy)-ethoxy]-ethyl} ammonium chloride. A powder containing 5% of this drug and ointments and tinctures with 1% active ingredient were marketed for the treatment of impetigo, ecthyma, dermatitis, infected eczema, dermatomycosis, and moniliasis as well as for mastitis in cattle. Between 1967 and 1972, about 24 kg of the antibiotic was made annually, of which about 5 kg was exported.[5] Manufacture ceased due to rising labor costs and decreasing abundance of the raw material, reindeer lichen (*Cladonia* sp.). About 40 kg of lichen is required to make 1 kg of usno which was claimed to be more effective than tyrothricin and bacitracin, and in several instances gave positive results even when tetracycline antibiotics were ineffective. Usnic acid was even claimed to have cured a case of tetanus which did not respond to treatment with the antiserum and more conventional antibiotics.[55] Usnic acid acts by uncoupling oxidative phosphorylation, and is effective because animal cells are far less permeable to this antibiotic than susceptible microorganisms. As with penicillin, a few people are allergic to usnic acid. Some of those affected react to +usnic acid; others only react to a mixture of +usnic and −usnic acid. The latter alone seems not to induce a response.[56]

In addition to the scientific use of lichen antibiotics, certain species, particularly *C. islandica*, are made into pastilles and teas. Decoctions of the latter are valued for their astringent flavor. The firm of F. Hunziker at Dietikon near Zürich uses 150 kg/year of lichen imported from the Balkan states to manufacture such products.[5]

Recently, there has been interest in another component of lichens, the polysaccharides.[57] Some possess antitumor activity and indeed have apparently been used traditionally against cancer.[58] In the case of sarcoma-180 in mice, it was found that a daily injection for 10 days of lichen extract 24 hr after implantation would lead to complete regression of the tumor in comparison with controls. In general, homoglucans with β (1 → 3) and (1 → 4) or (1 → 6) linkages were more effective than α glucans. Molecular weight and fine structure would also appear to be important as oat-lichenan and dextrans of various molecular weights are inactive.[59] The mechanism of action of these polysaccharides is not completely known, but is host mediated. The lichen polysaccharide extracts are thought to cause anoutpouring of lymphoid cells, plama cells, and macrophages in the vicinity of the grafted tumor.[60] Active polysaccharides have been found in genera such as *Umbilicaria, Lobaria, Usnea*, and *Sticta*.[61-63] Lichenan apparently has the side effect of inducing hardening and enlargement of the liver. Isolichenan, which also inhibits some cancers, does not cause this.[64] Other antitumor fractions have been extracted recently from *Ramalina almquistii* and found to be composed of d-protolichesterinic acid and nephrosterinic acid which were effective against the solid-type Erlich carcinoma.[65] These two lichen substances are examples of methylene lactones, a naturally occurring group of compounds with many other representatives that exhibit antitumor activity (e.g., vernolepin and elephantopin). Finally, 4-O-methylcryptochlorophaeic acid has proved to be an acidic nonsteroidal, antiinflammatory drug and has given clues to the mode of action of such drugs.[66] Clearly, these are interesting discoveries and at the very least will stimulate further work on the chemistry of lichens which, from the chemotaxonomic viewpoint, are probably the best-studied group of plants.[67]

VIII. CONCLUSIONS

The last 100 years have seen the decline of lichens as dye plants. It is perhaps only now that the scale of their former use is being fully appreciated as interest grows in industrial archaeology and delving into past lichenological records.[73]

The current prosperity in developed countries has resulted in unprecedented use of perfumes and lotions for which lichen-derived fixatives are so valuable. Until synthetic substitutes are found, a few lichen species will still be required on a large scale to fulfill this need and people will be involved in lichen picking on a commercial scale. It is to be hoped that the perfume industry and raw material producers will undertake research to ensure continued and optimal production via suitable woodland management and harvest rotation.

There is no doubt that lichens are used on a small scale for food, flavoring, and medicines in many parts of the world. Before fast foods and television change local customs, their uses must be documented and photographed from both a botanical and an ethnological viewpoint. Both Turner[21,22] and Lange[69] provide good models of this type of research. Lange investigated *Parmelia paraguariensis* which is used as a tobacco in Mauritania being imported from several hundred kilometers to the northwest where the plant grows. Turner in Canada has published on the food plants of the British Columbian Indians as well as on other aspects of their culture. These publications show that a study of the local uses of lichens (and indeed other plants) provides both a scientifically valuable record and fascinating reading. If this chapter stimulates one or more research students to undertake studies on the minor but interesting uses of lichens in particular countries, it will have served an important purpose in addition to reviewing the current situation regarding the major uses of these plants.

ACKNOWLEDGMENTS

I wish to thank Dr. M. R. D. Seaward, Miss V. Hinton, Dr. D. L. Hawksworth, and Mr. T. H. Moxham for making useful comments on the draft manuscript. I would also like to thank Prof. B. Fox and Prof. O. Lange for providing additional information. Dr. R. Honegger kindly collected for me several products made from lichens. Finally, I wish to express my appreciation to Dr. R. A. Donkin, Dr. S. Fregert, Mr. T. Moxham, Dr. M. R. D. Seaward, and Ms. E. Ziegenbein (Dragoco) for providing illustrations used in this chapter.

REFERENCES

1. **Brightman, F. H.,** Antibiotics from lichens, *Biol. Hum. Affairs,* 26, 1—5, 1960.
2. **Smith, A. L.,** *Lichens,* Cambridge University Press, London, 1920.
3. **Llano, G. A.,** Economic uses of lichens, *Econ. Bot.,* 2, 15—45, 1950.
4. **Llano, G. A.,** Utilization of lichens in the arctic and subarctic, *Econ. Bot.,* 10, 367—392, 1956.
5. **Richardson, D. H. S.,** *The Vanishing Lichens,* David & Charles, North Pomfret, Vt., 1975, 95—97.
6. **Moxham, T. H.,** Lichens in the perfume industry, *Dragoco Rep.,* 2, 31—39, 1981.
7. **Kauppi, M.,** The exploitation of *Cladonia stellaris* in Finland, *Lichenologist,* 11, 85—89, 1979.
8. **Berkow, R.,** *The Merck Manual,* Merck Sharp & Dohme Research Laboratories, Rathway, 1982, 312 and 2026—2034.
9. **Dahlquist, I. and Fregert, S.,** Contact allergy to atranorin in lichens and perfumes, *Contact Dermatitis,* 6, 111—119, 1980.
10. **Dahlquist, I. and Fregert, S.,** Patch testing with oakmoss extract, *Contact Dermatitis,* 8, 227, 1983.
11. **Thune, P. O., Solberg, Y. J., McFadden, N., Staerfelt, F., and Sandberg, M.,** Perfume allergy due to oakmoss and other lichens, *Contact Dermatitis,* 3, 396—400, 1982.
12. **Champion, R. H.,** Woodcutter's disease: contact sensitivity to lichen, *Br. J. Dermatol.,* 77, 285, 1965.
13. **Benezra, C., Ducombs, G., Sell, Y., and Foussereau, J.,** *Plant Contact Dermatitis,* Brian C. Decker, Toronto, 1985, 50—51
14. **Thune, P. O.,** Contact allergy due to lichens in patients with a history of photosensitivity, *Contact Dermatitis,* 3, 267—272, 1977.
15. **Thune, P. O. and Solberg, Y. J.,** Photosensitivity and allergy to aromatic lichen acids, composite oleoresins and other plant substances, *Contact Dermatitis,* 6, 64—71, 1980.
16. **Dalquist, I. and Fregert, S.,** Atranorin and oakmoss contact allergy, *Contact Dermatitis,* 7, 168—169, 1981.
17. **Richardson, D. H. S. and Young, C. M.,** Lichens and vertebrates, in *Lichen Ecology,* Seaward, M. R. D., Ed., Academic Press, London, 1977, 121—144.
18. **Hale, M. E.,** *The Biology of Lichens,* 3rd Ed., Edward Arnold, London, 1983, 138.
19. **Donkin, R. A.,** *Manna: An Historical Geography.* W. Junk, The Hague, 1980.
20. **Donkin, R. A.,** The "manna lichen": *Lecanora esculenta, Anthropos,* 76, 562—576, 1981.
21. **Turner, N. J.,** Economic importance of black tree lichen (*Bryoria fremontii*) to Indians of western North America, *Econ. Bot.,* 31, 461—470, 1977.
22. **Turner, N. J.,** *Food Plants of the British Columbia Indians, Part II, Interior Peoples,* British Columbia Provincial Museum, Victoria, 1978, 35—39.
23. **Brodo, I. M. and Hawksworth, D. L.,** *Alectoria* and allied genera in North America, *Opera Bot.,* 43, 1—164, 1977.
24. **Pulliainen, E.,** Nutritive values of some lichens used as food by reindeer in northeastern Lapland, *Ann. Zool. Fenn.,* 8, 385—389, 1971.
25. **Ingolfsdotter, K., Bloomfield, S. F., and Hylands, P. J.,** *In vitro* evaluation of the antimicrobial activity of lichen metabolities as potential preservatives, *Antimicrob. Agents Chemother.,* 28, 289—292, 1985.

26. **Gedner, C.,** *Miscellaneous Tracts Relating to Natural History, Husbandry and Physics*, Stillingfleet, B., Ed., London, 1759, 144—146.
27. **Scheffer, J.,** *The History of Lapland*, Translated by A. Cremer and printed for T. Newborough and R. Parker, London, 1674, 142.
28. **Henneguy, F.,** *Les Lichens Utiles*, Paris, 1883.
29. **Kobert, R.,** Ueber Giftstoffe der Flechten, *Sitzungsber. Naturforsch. Ges. Univ. Jurjew (Dorpat-Tartu)*, 10, 157—166, 1895.
30. **Schade, A.,** Über *Letharia vulpina* (L.) Vain, und ihre Vorkommen in der Alten Welt, *Ber. Bayer. Bot. Ges. Erforsch. Heim. Flora*, 30, 108—126, 1954.
31. **Slansky, F.,** Effect of the lichen chemicals atranorin and vulpinic acid upon feeding and growth of larvae of the yellow-striped armyworm, *Spodoptera ornithogalli, Environ. Entomol.*, 8, 865—868, 1979.
32. **Stephensen, N. L. and Rundel, P. W.,** Quantative variation and the ecological role of vulpinic acid and atranorin in the thallus of *Letaria vulpina, Biochem. Syst. Ecol.*, 7, 263—267, 1979.
33. **Gerson, U. and Seaward, M. R. D.,** Lichen-invertebrate associations, in *Lichen Ecology*, Seaward, M. R. D., Ed., Academic Press, London, 1977, 69—120.
34. **Lawrey, J. D.,** *Biology of Lichenized Fungi*, Praeger, New York, 1984, 232—247.
35. **Leechman, D.,** Aboriginal paints and dyes in Canada, *Trans. R. Soc. Can.*, Section II, 37—42, 1932.
36. **Poucher, W. A.,** *Perfumes, Cosmetics and Soaps*, Vol. 2, 8th ed., Chapman & Hall, London, 1975, 179—191.
37. **Moxham, T. H.,** The commercial exploitation of lichens for the perfume industry, in *Progress in Essential Oil Research*, Brunke, E. J., Ed., Walter de Gruyter, Berlin, 1986, 491—503.
38. **Serin, Y. K.,** Techno-economic Evaluation of Indigenous Lichens as Raw Material for Aromatic Resinoids, Annual Report, Regional Research Laboratory, Jammu-Kashmir, 1977, 51.
39. **Moxham, T. H.,** The use of lichen scrapers for gathering "oakmoss", *Br. Lichen Soc. Bull.*, 50, 18—19, 1981.
40. **Moxham, T. H.,** Lichens and perfume manufacture, *Br. Lichen Soc. Bull.*, 47, 1—2, 1980.
41. **Bergwein, I. K.,** Moss fragrance in modern perfumery, *Dragoco Rep.*, 3, 43—48, 1972.
42. **Bolton, E. M.,** *Lichens for Vegetable Dyeing*, Studio Press, London, 1960.
43. **McGrath, J. W.,** *Dyes from Lichens and Plants*, Van Nostrand Reinhold, Toronto, 1977.
44. **Casselman, K. L.,** *Craft of the Dyer: Color from Plants and Lichens from the Northeast*, University of Toronto Press, Toronto, 1980.
45. **Feddersen-Fieler, G.,** *Farben aus Flechten*, M & H Schaper, Hannover, 1982.
46. **Goodwin, J.,** *A Dyer's Manual*, Pelham Books, London, 1982, 87—91.
47. **Kok, A.,** A short history of orchil dyes, *Lichenologist*, 3, 248—271, 1966.
48. **Gordon, C.,** A petition of Dr. Cuthbert Gordon, *J. House Commons*, 41(305), 963—964, 1786.
49. **Gordon, C.,** *Memorial of Dr. Cuthbert Gordon relative to the discovery and use of cudber, and other dying wares, from the indigenous plants of this country. To which are added, certificates and documents, shewing the value of Dr. Gordon's discoveries to the woolen, cotton and linen manufacturers of Great Britain*, London, 1791.
50. **Henderson, A.,** Some memorabilia on the industrial manufacture of lichen dyestuffs, cudbear and orchil. II, *Bull. Br. Lichen Soc.*, 56, 22—24, 1985.
51. **Henderson, A.,** Some memorabilia on the industrial manufacture of lichen dyestuffs, cudbear and orchil, I, *Bull. Br. Lichen Soc.*, 55, 19—21, 1984.
52. **Moxham, T. H.,** Lichens and litmus, *Bull. Br. Lichen Soc.*, 50, 1—3, 1982.
53. **Hanssen, H. P. and Schadler, M.,** Pflanzen in der Traditionellen Chinesischen Medizin, *Dtsch Apoth. Ztg.*, 125, 1239—1243, 1985.
54. **Rajula, K., Outhen, O., Tuuponen, T., Lann, R., and Karkola, P.,** Pulmonary fibrosis with sarcoid granulomas and angitis associated with handling mouldy lichen, *Eur. J. Respir. Dis.*, 64, 625—629, 1983.
55. **Salo, H., Hannukela, M., and Hausen, R.,** Lichen picker's dermatitis [*Cladonia alpestria* (L.) Nabenh.], *Contact Dermatitis*, 7, 9—13, 1981.
56. **Heki, M., Nishikawa, T., and Fugii, M.,** Experminental studies in the treatment of tetanus and on a case of tetanus treated with USNIC acid, *Jpn. J. Med. Sci. Biol.*, 5, 98—100, 1952.
57. **Gorin, P. A. J. and Iacomini, M.,** Polysaccharides of the lichens *Cetraria islandica* and *Ramalina usnea, Carbohydr. Res.*, 128, 119—132, 1984.
58. **Hartwell, J. L.,** Plants used against cancer. A survey, *Lloydia*, 34, 386—438, 1971.
59. **Shibata, S.,** Polysaccharides of lichens, *J. Natl. Sci. Counc. Sri Lanka*, 1, 183—188, 1973.
60. **Tokuzen, R.,** Comparison of local cellular reaction to tumor grafts in mice treated wit some plant polysaccharides, *Cancer Res.*, 31, 1590—1593, 1971.
61. **Nishikawa, Y, Yanaka, M., Shibata, S., and Fukuoka, F.,** Polysaccharides of lichens and fungi. IV. Antitumour active O-acetylated pustulan-type glucans from the lichens of *Umbilicaria* species, *Chem. Pharm. Bull. Jpn.*, 18, 1431—1434, 1970.

62. **Nishikawa, Y., Ohki, K., Takahashi, K., Kurono, G., Fukuoka, F., and Emori, M.,** Studies on the water-soluble constituents of lichens. II. Antitumor polysaccharides of *Lasallia, Usnea* and *Cladonia* species, *Chem. Pharm. Bull. Jpn.*, 21, 1014—1019, 1974.
63. **Takahashi, K., Takeda, T., Shibata, S., Inomata, M., and Fukuoka, F.,** Polysaccharides of lichens and fungi. IV. Antitumour active polysaccharides of lichens in the Stictaceae, *Chem. Pharm. Bull. Jpn.*, 22, 404—408, 1974.
64. **Fox, B.,** Lichens in medicine, *Bull. Br. Lichen Soc.*, 54, 4, 1984.
65. **Hirayama, T., Fujikawa, F., Kashara, T., Otsuka, M., Nishjida, N., and Mizuno, D.,** Anti-tumor activities of some lichen products and their degradation products, *Yakugaku Zasshi*, 100, 755—759, 1980.
66. **Shibuya, M., Ebizuka, Y., Noguchi, H., Iitaka, Y., and Sankawa, U.,** Inhibition of prostaglandin biosynthesis by 4-0-methylcryptochlorophaeic acid; synthesis of monomeric arylcarboxylic acid inhibitory activity testing and X-ray analysis of 4-O-methylcryptochlorophaeic acid, *Chem. Pharm. Bull. Jpn.*, 31, 407—413, 1983.
67. **Culberson, C. F., Culberson, W. L., and Johnson, A.,** *Chemical and Botanical Guide to Lichen Products*, 2nd Suppl., American Bryological and Lichenological Society, St. Louis, Mo., 1977.
68. **Elix, J. A., Whitton, A. A., and Sargent, M. V.,** Recent progress in the chemistry of lichen substances, *Prog. Chem. Org. Nat. Prod.*, 45, 104—234, 1984.
69. **Lange, O. L.,** Die Flechte *Parmelia paraguariensis* als Handelsware in der Südlichen Sahara, *Nat. Volk.*, 87, 266—273, 1957.
70. **Fiell, K. M.,** personal communication.
71. **Hawsworth, D. L.,** personal communication.
72. **Seaward, M. R. D.,** personal communication.
73. **Perkins, P.,** Ecology, beauty, profits: trade in lichen-based dyestuffs through western history, *J. Soc. Dye Colourists*, 102, 221—228, 1986.

Chapter XII.C

LICHENS AND PEDOGENESIS

David Jones

I. INTRODUCTION

That saxicolous (rock-inhabiting) lichens affect their substrate has been recognized for some time. Thus, we find in a report on studies by Sollas[1] that minute hemispherical pits sprinkled over the surface of many exposed limestone faces were produced by apothecia of *Verrucaria rupestris*. It was referred to as a "case of excavation by purely chemical action," and showed that the action of lichens "is not purely conservative".

Much controversy has arisen over the years regarding the ability of saxicolous lichens to degrade their rock substrates, thus leading to the first stages in the formation of soil. The general conclusion of Cooper and Rudolph[2] in their assessment of studies by various researchers was that the "classical role of lichens in soil formation and plant succession in rocky places is questioned", that "the importance of lichens has been exaggerated and that the developmental story of the vegetation has been oversimplified." On the positive role of lichens in soil formation, Emerson (in Cooper and Rudolph[2]) commented that "the crustose forms are the world's great pioneers" and that the mycobiont (the fungal partner of the fungus-alga symbiosis) mycelia penetrate the rock crevices, gradually breaking loose small particles, with secretions from the cells of the lichen dissolving parts of the stone. Clements and Shelford[3] stated that "the initial conversion of rock into soil is carried on by the pioneer lichens and their successors, the mosses, in which the hair-like rhizoids assume the role of roots in breaking down the surface into a fine dust". Cooper and Rudolph[2] also referred to a number of papers from the early part of the 20th century which concluded that lichens cause some damage to certain rock surfaces, although the amount is comparatively slight. Schatz[4] was of the opinion that "geology, agriculture, and botany have yet to accord lichens the respect they merit" because as well as being important weathering and pedogenic agents, they may also be important sources of soil nitrogen which is released on their decay. The author also stated that the importance of lichens in soil formation, and the implication of lichen acids in chelation, brought about the development of the agricultural use of chelating agents to enhance the growth of higher plants.

The biological weathering of rocks and their constituent minerals involves both biogeochemical and biogeophysical weathering processes. The latter have been defined by Silverman[5] as "those processes by which life forms cause mechanical fracturing and disruption of rocks and minerals to produce particles smaller than the original material". The same author defined biogeochemical weathering as referring to "all other processes, direct or indirect, by which living organisms and their metabolic processes and products affect the chemical stability and composition of silicate rocks and minerals". Both these aspects of weathering are referred to in this review, explaining the role of lichens in these processes.

The weathering effects of lichens on rock surfaces can manifest themselves as etch markings on the constituent rock minerals and can involve the conversion of certain minerals to siliceous relics, the precipitation of poorly ordered aluminous ferruginous material on the rock surface beneath the lichen, and the formation of crystalline organic salts in the lichen thallus. All these features can now be studied by electron microscopic techniques[6,7] combined with analytic techniques such as X-ray diffractometry and IR spectrometry.

II. BIOGEOPHYSICAL AND BIOGEOCHEMICAL WEATHERING: MECHANISMS AND EFFECTS ON SUBSTRATES

In an early study of the mechanical action of crustose lichens on shale, schist, gneiss, limestone, and obsidian, Fry[8] noted that "when removed from their rock substrates, both dark and light hypothalli show mineral particles attached to, or embedded among, the hyphae". However, the techniques to study surfaces of these particles were not available at that time. In fact, the limitations imposed by the magnification of the microscope lenses used were referred to in connection with the location of small mineral fragments embedded in, or beneath, lichen thalli. The object of the study was to show that some of the disintegration of the superficial layers of rock substrates was due to the expansion and contraction of the thalli of the epilithic crustose lichens (and these included *Lecidea confluens, L. plana, Lecanora sordida, L. atra, L. sulphurea, L. parella, Rhizocarpon geographicum, Aspicilia alpina,* and *A. calcarea*) firmly and closely attached to the substrate. Although the author specifically stated that the chemical action of these lichens on the substrate was not being discussed, some interesting observations were made in this connection. Thus, reference is made to "mineral fragments below the two fused apothecia are brown, different chemically from, and softer than, the rest of the schist". There is a contradiction in the following page of the paper where a statement is made that "although in contact with the lichen for some long time, yet, as far as one can detect with the microscope, there appears no chemical alteration in the minerals, even in those parts directly in contact with the purple-stained hyphal walls". The author concluded that chemical decomposition of the rock minerals by means at the disposal of the lichens is a very slow process and also that in schists and shales their mechanical disintegration precedes chemical decomposition of the minerals. The organism under study in that particular part of the work was *L. sordida*. Beneath the apothecia of *R. geographicum* on schist very few minerals could be traced. Evidence for biogeochemical weathering of rock surfaces was given, however, in the case of *Aspicilia*, and *R. geographicum* colonizing obsidian, a hard acid rock in the nature of glass. Very significantly, the author found that when the lichen was removed from the polished obsidian surface, the latter was seen to be etched by very minute circular or elongated depressions. The suggestion was made that, in this instance, the action was chemical since the etching was distributed evenly below older lichens, but was absent on the rock over which the younger thalli were growing. In spite of this statement, the author still had doubts since the suggestion was made that, indirectly, the etching effect might owe its origin to the contraction of the lichen tissue followed by the sudden separation of the hard carbonaceous material from the substrate. Evidence for this was provided by data on the action of artificial apothecia (made with gelatin) on rock surfaces.

In a previous publication on a suggested explanation of the mechanical action of lithophytic lichens on rocks (shale), the same author (Fry[9]) concluded, from experiments with gelatin and glass to simulate the effect of drying out of the lichen on its substrate, that the initial stage of the mechanical disintegration of the altered rock by the lichen thalli (in this case, *Xanthoria parietina*) was followed by the chemical decomposition of the separated fragments. No evidence, however, was provided for the latter process.

Earlier papers certainly indicated the involvement of biological processes, more by physical means, in the weathering of mineral substrates, but these researches were seriously hampered by the lack of good optical and, more particularly, scanning electron microscope facilities. In addition, they did not consider the type of chemical reaction that would result in the degradation of the mineral substrate.

No significant progress in working out the mechanisms involved in the weathering of rock minerals by lichens seemed to have taken place during the 20 years after Fry's observations on lichens and experiments with gelatin drying on glass surfaces to explain the mechanical forces of lichens on their substrates. Jacks[10] has published a useful collection of notes of work done by Russian researchers in the field of organic weathering. Thus, Polynov[11] involved himself in the study of the first stages of soil formation on massive crystalline rocks. The activity of lichens resulted in the mechanical destruction of the rock and also, more significantly, was claimed to effect a biochemical transformation of its mineral constituents resulting in new minerals, in particular quartz and aluminosilicates of the montmorillonite type in which aluminum is replaced by magnesium in the crystal lattice. Jacks also referred to the work of Bobritskaya,[12] which indicated that the nature of the rock influences the generic nature of the lichens and that this in turn determines the composition of the ash and the quality of the organomineral fine earth subsequently produced. Thus, on acid rocks *Parmelia* predominated, on basic silicate rocks *X. parietina* was dominant, and on calcareous rocks only *Acarospora* and *Lecanora crassa* existed.

A. Role of Lichen Acids in Biogeochemical Weathering

Schatz et al.[13] and Schatz,[14] in their studies on chelation as a biological weathering factor in pedogenesis, recognized lichens as being convenient material for investigating certain problems at a simple level because they represent "an early unstabilized stage in evolution". They drew attention to the fact that lichens contain unusually large amounts of a wide variety of complex organic compounds not found elsewhere in nature and that many of these are powerful chelators by means of which lichens weather rock material and, furthermore, extract essential trace elements. They also suggested that chelation is a major biological weathering factor qualitatively distinct from acidity. Initially, Schatz et al.[13] tested the weathering action of carbonic acid on limestone, gneiss, and shale. Also tested were citric acid and ethylenediaminetetracetic acid and these two proved to be active weathering agents. A number of other acids (including hydroxy, keto, and amino aliphatic acids) were also included for comparison and all tested in the alkaline range. The minerals were attacked to varying extents and, since the observed weathering occurred at pH 8.0, it was not a simple acid digestion.

Schatz[14] followed up this work to obtain information about weathering effects of mixtures of lichen constituents on different minerals. In addition, the interaction of two lichen acids with rock material was examined. Ground, dry lichen was mixed with the minerals in an aqueous solution. Supernatants were analyzed spectrophotometrically. Two specimens of a saxicolous lichen, *Parmelia conspersa*, "reacted well" with granite and mica in a matter of hours resulting in a rise in pH and the development of a distinct reddish color. *Umbilicaria arctica* also reacted with granite but was not tested on mica and marl. *Caloplaca elegans*, although a rock lichen, did not bring granite or mica into solution. It was interesting that the most rapid reaction occurred with granite and the specimen of *P. conspersa* which was growing on granite when collected. These saxicolous lichens are known to contain the lichen acids parietin, salacinic, and gyrophoric. The results obtained on incubating minerals with pure lichen acids showed that physodic acid was more reactive than lobaric acid on both granite and mica, and it was concluded that the low solubility of lichen acids does not prevent these substances from actively weathering rocks and minerals via chelation. Schatz considered that the rapid reaction between *P. conspersa* and granite, for example, was at variance with Hale's[15] view that although "lichens probably do take some part in soil formation, their effects must be measured in terms of centuries, not decades". Schatz emphasized that the hyphae of crustose saxicolous lichens are in intimate contact with mineral particles (as is also the case in many mycorrhizas) and that the lichen acids are extracellular, crystallizing out on hyphae where they can act directly on rock minerals.

Further evidence for the involvement of lichen acids in the weathering of rock minerals was presented by Syers,[16] who studied the interaction of fumarprotocetraric acid, a lichen depsidone, and *P. conspersa* (in ground, dried form), a saxicolous lichen, with granite using a spectrophotometric method. The results obtained indicated that there was a considerable reaction between granite and both lichen acid and lichen. It was noted that fumarprotocetraric acid is found in *P. conspersa* and that the lichen was frequently found growing on granite. The author concluded that the results supported the findings of Schatz,[14] which suggested that lichen acids are sufficiently soluble to function as chelating agents. In this context, the work of Dawson et al.[17] is of relevance since they found lichen compounds such as usnic acid and other acids at different levels in soil profiles in arctic Alaska which were colonized by the indigenous lichen *Cladonia mitis*. They suggested that these acids may contribute significantly to the development of such profiles.

In an examination of the impact of lichens on podzolization in arctic Alaska, they identified the acid products of *C. mitis* and determined the extent of their natural mobilities, and also measured their migration in the soil profile. Their findings suggested an important role for lichens in the formation of podzolic soils in the conifer woodland and in the alpine tundra of the Brooks Range. Several lichen compounds common to *C. mitis*, including usnic acid, atranorin, rangiformic acid, and psoromic acid, were found in the soil solutions as well as within the soil profile.

The vegetation cover of the soils analyzed were lichens, mixed spruce/lichens, and spruce. The lichen compounds, and in particular usnic acid, were found not to be absorbed in the B21hir horizon and thus are mobile, penetrating at depth into the soil profile. The authors also referred to the known antibiotic characteristics of usnic acid which may thus be important in selecting or inhibiting the soil microflora.

Iskandar and Syers[18] published a later paper on metal-complex formation by lichen compounds. Six such compounds (generally referred to as lichen acids, although not all lichen organic compounds are acids) were tested for their reaction with water suspensions of biotite, granite, and basalt. The formation of soluble, frequently colored, complexes indicated that chemical weathering had occurred. The acids released greater amounts of Ca than Mg, Fe, and Al from the silicates, and the release of Ca was usually greater from biotite than from granite or basalt. As anticipated, citric, salicylic, and other organic acids used as controls, because of their higher water solubility, released greater amounts of cations from the silicates than did the lichen acids. The conclusion was that lichen compounds are sufficiently soluble in water to form soluble metal complexes and to effect chemical weathering of minerals and rocks.

Williams and Rudolph[19] pursued the chelation theme by determining the iron chelating ability of certain lichen symbionts and free-living fungi isolated from sandstone rock surfaces. The iron chelating ability of *C. squamosa* was also considered by using squamatic acid, its sole lichen compound. The ability to chelate was interpreted as a form of biochemical weathering. Their results showed that the myco- and phycobionts of *Caloplaca holocarpa* and *L. dispersa*, both collected from a limestone wall, and of *Cladonia cristatella* and *C. squamosa* (phycobiont not isolated) from sandstone outcrops, failed to chelate iron (ferric sulfate) as substrate in culture. This supported the premise that it is the lichen acids, formed only by the whole lichen, and not any metabolic compounds of either of the symbionts, which are responsible for any metallic binding. In fact, it would appear that the production of a lichen acid by either the myco- or phycobiont has never been established in axenic culture. The authors were, however, able to demonstrate that squamatic acid, extracted by them from *C. squamosa*, solubilized ferric iron from ground and sterilized sandstone. Schatz et al.,[20] in an attempt to obtain information on the significance of chelation phenomena in pedogenesis and soil fertility, also studied the chelating ability of lichen acids on various rock minerals. They made pH measurements on mixtures of ground, dried lichen thalli, five of the species from soil habitats, and a range of mineral materials including limestone, rock phosphate, granite, basalt, and gypsum. Changes in pH, generally associated with chelation phenomena, occurred in all but one lichen, the mixtures being acid. Pure specimens of lichen acids were not studied. The suggestion was made that "with respect to higher plants, it might be possible to breed and select strains that secrete greater amounts or different kinds of chelating agents''. Thus, such plants could adapt to areas where nutrient minerals, derived from rocks, are limiting.

Ascaso, et al.[21] were probably the first team to investigate the nature and composition of the mineral content of lichens scraped from rock surfaces. They used a combination of analytical methods which included X-ray powder diffraction techniques, atomic absorption spectrophotometry, and flame photometry. They also used transmission electron microscopy to examine minerals in the scrapings after removal of the organic matter with hydrogen peroxide. Thus, examination of such material from the thallus of *P. conspersa* colonizing granite revealed kaolinite, halloysite, and amorphous silica, and when this lichen was associated with gneiss, only the latter two were detected. Of significance was the finding, from electron microscopic techniques, that the surfaces of micas and chlorites were etched in some samples. Similarly, the micas and feldspars showed alteration (weathering) at the interface between *R. geographicum* and its granite substrate and halloysite was the most frequent newly formed mineral. In contrast, no weathering of the rock minerals could be detected under the thalli of *U. pustulata*. When a mixture of four lichen compounds — atranorin, usnic acid, stictic acid, and norstictic acid — was incubated with either rocks or minerals, which included granite, gneiss, albite, orthoclase, biotite, muscovite, and quartz, the results were essentially the same as those obtained by using stictic acid alone, thus confirming that this acid was the most active in the process of rock and mineral weathering. The overall conclusion of these authors based on the above information, as well as that from experiments incubating cleaned dried and crushed water extracts of the lichen thalli with various minerals, was that lichens are able to induce both chemical and morphological changes in rocks as well as in their mineral content. This was the first comprehensive study of the reaction of lichens with their mineral substrates and the first attempt to explain and interpret the weathered features of the individual mineral particles, observed by transmission electron microscopic techniques, and also relate them to specific lichen acids.

B. Role of Oxalic Acid in Biogeochemical Weathering

The significance of oxalates found at rock/lichen interfaces and the possible implication of oxalic acid in weathering rock minerals have received much attention in recent years. In 1980, Jones et al.[22] published their first paper in this context. They studied the weathering phenomena resulting from the encrustation of basalt by the lichen *Pertusaria corallina* using electron microscopic techniques and other methods including X-ray powder diffractometry, electron diffraction, and IR spectroscopy. All these techniques were invaluable in interpreting and explaining the observed phenomena. Thus, it was shown that the growth of the lichen on the rock, rich in plagioclase feldspar and ferromagnesian minerals, resulted in extensive etching of the primary rock-forming minerals and in the degradation of the clay minerals to yield a thin ochreous crust of ferruginous (ferrihydrite) and alumino silicate materials. It was proved from pure culture studies with the mycobiont, isolated on agar, and a free-living fungus used to effect experimental mineral weathering, that the weathering patterns were brought about principally by the oxalic acid secreted by the mycobiont. Thus, this oxalic acid was probably instrumental in dissolving the calcium and other elements from the rock minerals, although the calcium oxalate was for the most part crystallized as an insoluble extracellular precipitate (Figure 1). This work was followed by a study, employing the same techniques, of the weathering phenomena brought about by the growth of *L. atra* on serpentinite, which consists mainly of magnesium silicate minerals.[23] Again the authors deduced from similar experimentation that the mycobiont was responsible for the degradation of the magnesium silicates and, in particular, chrysotile, by secreting oxalic acid. This resulted in the decomposition of appreciable amounts of crystalline magnesium oxalate dihydrate at the rock-lichen interface (Figure 2). The only weathering product detected in the lichen weath-

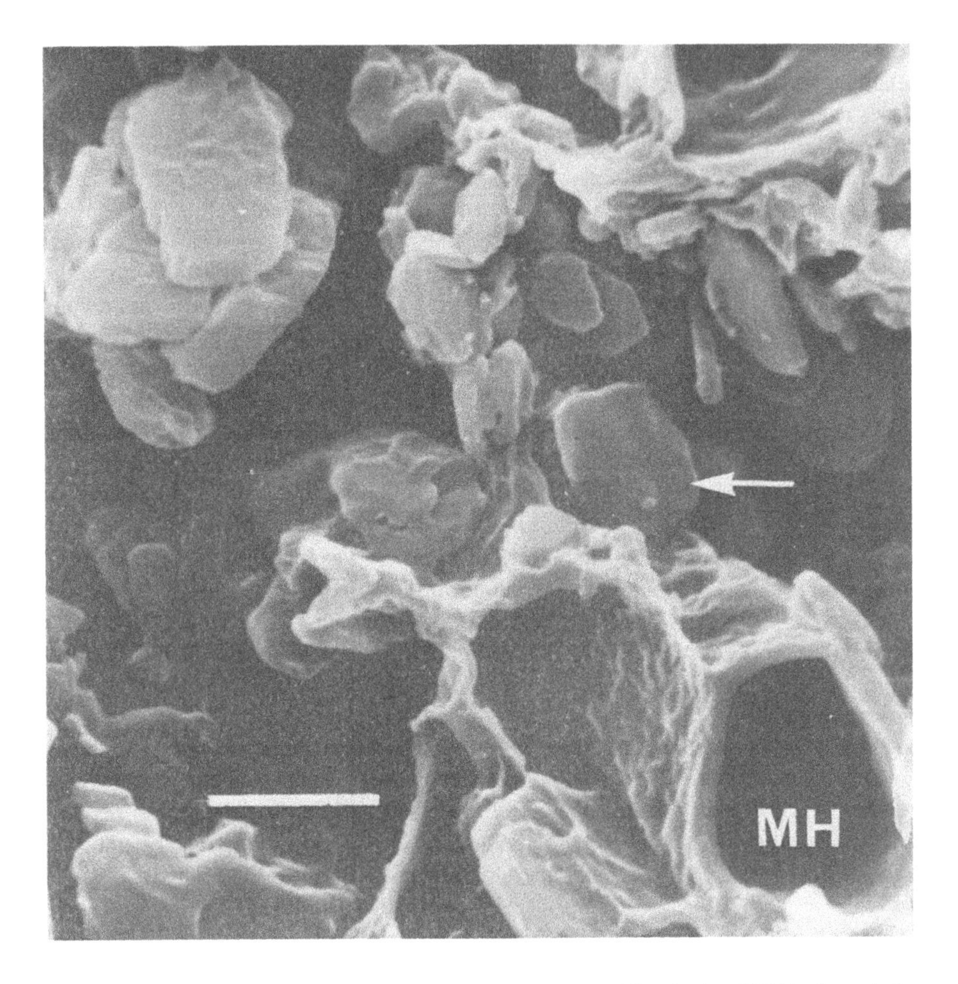

FIGURE 1. Calcium oxalate crystals (arrowed) associated with mycobiont hyphae (MH) in the saxicolous lichen *Pertusaria corallina* from Uist, Outer Hebrides, Scotland. (Scale bar = 2.5 μm.)

ering crust was an X-ray amorphous silica gel which often retained the fibrous morphology of the chrysotile from which it is formed. Scanning electron microscopy of the minerals found embedded in the lichen thallus (largely fibrous chrysotile and nonfibrous antigorite or lizardite, magnetite, and an unidentified iron silicate) revealed decomposition, although to varying degrees. Deep pitting occurred on the generally fragmented magnetite and iron silicate. Chrysotile fibers on the rock surface on which the lichen had been growing revealed fibers with a split and twisted appearance with evidence of an associated gel-like material. Electron probe microanalysis of these fibers yielded an X-ray spectrum dominated by silicon. Magnesium and silicon counts taken from the chrysotile in thin section at various depths below the rock-lichen interface showed conclusively that the chrysotile was depleted in magnesium up to 100 μm below the interface.

A further publication by Wilson and Jones[6] described the interaction of the lichen *Pertusaria corallina* with its substrate rock, in this case manganese ore which consisted of mixtures of powdery lithiophorite ([Al, Li] Mn O_2 $[OH]_2$) and hard cryptomelane (KMn_8O_{16}; this also contained hollandite — Ba Mn_8O_{16}). Although in this case the interface between the lichen and cryptomelane showed little indication of a loose weathering crust, crystals were detected in the lichen thallus which were subsequently isolated and shown to be manganese oxalate dihydrate. The interpretation of this finding was that, once more, oxalic acid was implicated in the reaction of the lichen with the substrate rock.

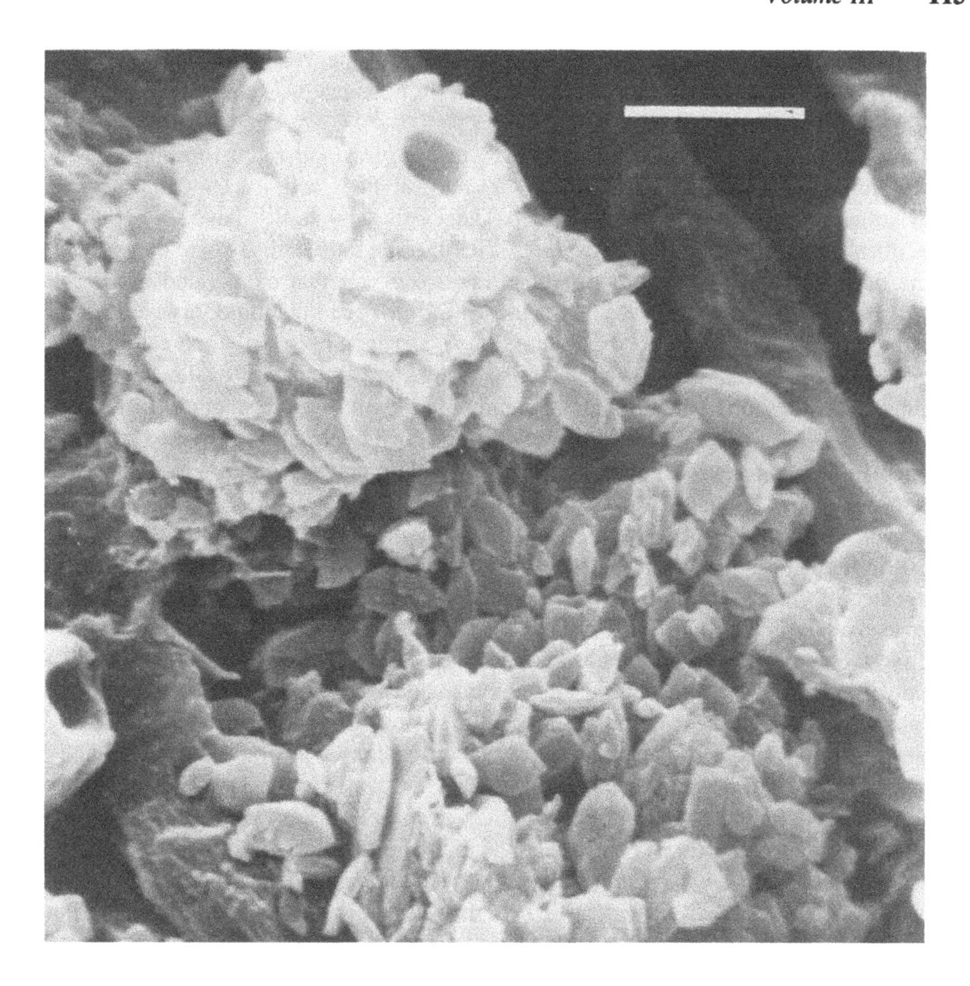

FIGURE 2. Magnesium and calcium oxalate crystals encrusting mycobiont hyphae in the saxicolous lichen *Lecanora atra* on serpentinite, Leslie, Aberdeenshire, Scotland. (Scale bar = 5 μm.)

Two further papers on lichen weathering of minerals, one dealing with implications for pedogenesis and another on the application of scanning electron microscopy (SEM) and microprobe analysis to such studies, have been published recently.[7,24] The conclusion was that the weathering effects from crustose lichens are similar to those in soils developed under cool temperature climates. In both instances, there is a breakdown of primary minerals and SEM observations show signs of intensive etching of primary minerals in the sand fraction of Scottish soils.

Ascaso et al.[25] have reported that in the alteration of the particular rock which they studied, the most active lichens were those whose thallus adheres best to the substrate but that all the species, *Caloplaca callopisma, Diploschistes ocellatus, Squamarina oleosa*, and *Problastenia testacea*, had an intense effect in destroying the primary minerals of the rock which consisted of calcite or dolomite. Of significance, also, was their observation that the weathering by *D. ocellatus* leads to the formation of significant amounts of calcium oxalate monohydrate (whewellite). This compound was also present in the other interfaces, although in smaller amounts. On the rocks rich in Fe, *C. callopisma* formed considerable amounts of ferric oxalate. The authors concluded that oxalic acid brought about the alteration of the rock. In the same year, a paper was published by Vidrich et al.[26] on the activity of lichens (not identified) on rocks in a "natural Tuscan environment". They concluded that the lichens brought about the dissolution of primary minerals and not the formation of neogenesis minerals. Large amounts of calcium oxalate were found in the weathered sheets of calcareous rocks, while in the ferromagnesian rocks, soluble iron oxalate was easily washed away and this facilitated further weathering.

Since lichens can be regarded as primitive mycorrhiza (soil fungi forming a symbiotic relationship with roots of higher plants), a study of their activity on rock surfaces has implications for mineral weathering in soil. Thus, Graustein et al.[27] have shown the occurrence of calcium oxalate crystals in the litter layer of some forest soils, where they accumulate on the outer surfaces of the fungal hyphae. This suggests that oxalic acid is secreted by them. Cromack et al.[28] have presented evidence for the accumulation of crystalline calcium oxalate in the fungal mats of the basidiomycete *Hysterangium crassum*, an ectomycorrhizal fungus of Douglas fir (*Pseudotsuga menziesii*). The large amounts of oxalic acid exuded by the fungus were responsible for the precipitation of calcium in the form of crystals of calcium oxalate. SEM showed the dense network of mycorrhizal hyphae to be covered by weddelite crystals. In addition, large hyphae were shown to be in contact with grains of andesite, an igneous rock type common in the parent material of the soils under study. The authors inferred that oxalate (= oxalic acid) was exuded by the fungal hyphae and reacted with Fe and Al in the andesite, thus accelerating weathering. They illustrate the pitted and cracked appearance of the andesite grains which suggested intense chemical weathering.

Since oxalate anions and oxalic acid are chelators of iron and aluminum, they could improve the availability of soil phosphate to plants, because phosphate is known to form insoluble compounds with iron and aluminium oxides. Calcium oxalate, as stated above, can accumulate on surfaces of fungal mycelium in soil, can appear as spines on spores of certain soil fungi, and also can occur as crystals in many plants and thus subsequently, on the decay of the latter, will be incorporated into the soil. In fact, oxalate production by ectomycorrhizal fungi at the root surface may explain increased rates of weathering and nutrient release by mycorrhizal compared to nonmycorrhizal roots.[29] The mycorrhizas studied were associated with *Pinus radiata* and *Eucalyptus marginata* in Western Australia; the mycorrhiza of *P. radiata* was identified as that formed by *Rhizopogon luteolus* but the symbiont of *E. marginata* could not be identified.

There have been some studies in recent years where the phenomenon of lichen weathering of their rock substrates has been established, but without supporting evidence for the implication of lichen acids and organic acids such as oxalic. However, corrosion of the minerals has been recorded, largely by using electron microscopes, so some reference is made to these observations. Thus, Hallbauer and Jahns,[30] in the course of geochemical investigations of lichens, and using a scanning electron microscope, observed quartz grains incorporated in the thallus of the lichen *Dimelaena oreina* which were pitted. The lichen in this instance was colonizing quartzites and quartzitic conglomerates of a waste rock dump from a gold mine in South Africa. Their microscope was fitted with an energy dispersive X-ray microanalyzer, and consequently the authors were able to analyze the elemental composition of both the mineral particles and the mycobiont hyphae which appeared to penetrate the quartz creating the impression of chemical boring. Thus, we have what was probably the first preliminary account of lichen weathering of its mineral substrate where a scanning electron microscope fitted with a microanalyzer was used. An earlier detailed study, employing the scanning electron microscope, was made on carbonate rocks colonized by lichens as well as algae and fungi,[31] but no microanalysis was used. This work showed that endolithic lichens produced the deepest rock alteration. Such lichens grow from the surface into the rock, leaving a portion of the vegetative thallus or reproductive structure on the rock surface.

Golubic et al.[32] differentiate endolithic organisms into three categories: chasmoendoliths if they inhabit fissures in rocks; cryptoendoliths if they dwell within structural cavities, and euendoliths if they actively penetrate calcareous substrates. They used this terminology to describe the ecological niches of microorganisms within hard mineral substrates; the organisms which were attached to the external surfaces were referred to as epiliths. They regarded all organisms that inhabit hard rock substrates as lithobionts.

One of the most interesting papers on the activity of cryptoendolithic lichens is by Friedmann,[33] who described their occurrence in sandstones in the frigid desert of the Antarctic dry valleys where there are no visible life forms on the rock surface. In these lichens, ''loose filaments and cell clusters grow between and around the crystals of the rock substrate so that the lichen is embedded in the rock matrix, covered by the hard surface crust''. The loose filaments are the mycobiont hyphae and are often covered with precipitated iron compounds in rocks where iron solubilization takes place. The end result of the activity of the lichen was a characteristic exfoliative weathering pattern and in the iron-rich dark sandstone resulted in a colorful patchwork pattern of light brownish surface crust and dark reddish-brown exposed rock. The comment was made that the cementing substance between the sandstone grains is apparently solubilized at the level of the lichen. The relationship of the lichen with its rock substrate was excellently illustrated by the authors, using both light microscopy of fractured rock surfaces and scanning and transmission electron microscopy (SEM and TEM) of the phyco- and mycobionts. The microscopical details were supported by a chemical analysis of metal concentrations in the different layers of colonized sandstone and bed rock.

Galvan et al.,[34] in a study on the pedogenic action on metamorphic rocks, concentrated on the lichen-rock interface, using SEM and TEM, coupled with IR spectroscopy, and X-ray diffraction techniques. They examined the nature and composition of the minerals present at the interface between four lichens — *Parmelia conspersa, Parmelina tiliacea, Lasallia pustulata*, and *Ramalina protecta* — and several metamorphic rocks including schists and gneiss. Their conclusions were that the lichen species were able to form gels between the thalli and rocks and that micaceous minerals are those most often retained under the thallus. They also concluded that there is a direct relationship between the amount of the extracted minerals and the rock texture.

III. PLANT SUCCESSION AND SOIL DEVELOPMENT

A. Plant Succession

Ugolini and Edmonds[35] considered that the pedogenic significance of lichens is largely restricted to rock faces and cliffs and is of prominence in harsh environments such as Antarctica. They concluded that lichens established on bare rocks trap atmospheric dust and other debris which, when mixed with the lichen thalli, contributes to the formation of primitive soils.

Among the first plants to colonize bare rock outcrops are crustose lichens, but there are exceptions as pointed out by Oosting and Anderson,[36] who studied the vegetation of a bare cliff face in western North Carolina. They noted that in the early stages of mat formation, the mosses *Rhacomitrium heterostichum ramulosum* or *Andreaea rupestris*, the initial colonizers of the rock, were subsequently ''invaded'' by two lichens, *Cladonia subcariosa* and *C. coccifera*, which grew on them, or between them, wherever soil particles or organic matter accumulated. Other *Cladonia* spp. also occurred, particularly on mats with large amounts of organic matter, and the suggestion was made that the water-holding capacity of the mat was increased thus making it possible for further colonization by other plant species. The collection of soil and other materials was thus greatly increased under these conditions so that other species of mosses could colonize the substrate which in turn greatly facilitates the collection of soil. The grasses *Danthonia spicata* and *Panicum huachucae* subsequently established themselves on these mats. Finally, when a mat of several inches had built up, the first woody species, *Chionanthus virginica*, made its appearance and later seedlings of *Acer rubrum* gained a foothold.

Keever et al.[37] have also made an interesting study of the plant succession on exposed granitic rock in North Carolina. The pioneer colonizers on the hot dry rocks (bare rock phase) were crustose lichens which were able to withstand desiccation and grow without soil. The most prominent was *Staurothele diffratella*, but *Peccania kansana* also was found. The authors commented that these lichens could have had "some corrosive effect on the granite, but there is no evidence that it is enough to be a factor in permitting other plants to follow". *Parmelia conspersa*, a foliose lichen, also grew on the bare rock, but there was no evidence to show that it contributed to mat building. In fact, the first pioneer that showed evidence of soil building and consequent succession was the moss *Grimmia laevigata*, and sometimes *G. pilifera*.

In glaciated countries, the vegetation growing on rock is often prominent and also is of great ecological interest.[38] Because of the virtually complete absence of soil in the early stages of colonization, lichens and mosses often play an important role in forming "cushions" which harbor humus and mineral particles. On strongly acid rock, the author described a typical succession of crustaceous or umbilicariaceous lichens followed by *Parmelia* spp., ending in *Cladonia* spp. growing on a thin humus layer. As the soil layer grows thicker, mosses become more common. On soil covered by the lichens, further lichens, including *Baecomyces rufus* and/or *Lecidea humosa*, may be found.

B. Nutrient Accumulation

Bobritskaya[12] has published a lengthy account of the absorption of mineral elements from massive crystalline rocks by lichens and mosses in an attempt to determine how soil originates and to ascertain the process by which it is formed from the barren rock. Initially, a useful review of the accumulation of elements such as phosphorous, sulfur, and potassium in lichens is given, together with some discussion on the role of lichens in weathering processes. The work of Yarilova[39] on the ability of *Haematomma ventosum*, *Squamaria rubina*, and *Gyrophora cylindrica* to concentrate S, P, and K in the high mountain regions of the North Caucasus was referred to by Bobritskaya, and the comment was made that this concentration of elements is not an exceptional occurrence or a local phenomenon. The study of Bobritskaya[12] was to determine the ash composition of the flora (including lichens) on rocks, particularly granite, but also limestone, which frequently participates in the formation of primary soils. A specific sequence for colonizing the rock surface existed, the first lichens to appear being the crustose forms, followed by the foliose and bushy lichens. Subsequently, mosses colonized the cavities "excavated" in the rock surface by the lichens. *Acarospora* and *Lecanora crassa* were specific to limestone, whereas *Parmelia* colonized acid igneous and intermediate rocks, and *Xanthoria parietina* favored basic silicate rocks and limestone. In the case of *Gyrophora*, a concentration of potassium was noted and all representatives of the primary lichen colonizers concentrated phosphorus and sulfur. One of the concluding remarks made was that, in the particular system studied, the primary flora, including lichens, constitutes the material from which the organic part of primary soil is evolved and, in the mineralization of this component, the elements for the mineral nutrition of higher plants are released. The ability of these primary soils developing on sheer rocks to support higher plants was thus due to the ash composition of the primary flora, which contains a high content of P, S, and K, and also the presence in adequate quantities of other elements of ash nutrition such as Ca, Mg, and Fe. Thus, the primary flora was able to provide the most important elements of fertility in primative soils. As it is difficult to separate out mineral particles from lichen thalli, the above types of analyses could include mineral and organic forms of the elements determined. This is a difficulty that needs to be taken into account in the interpretation of these data.

Syers and Iskandar[40] in their review on the pedogenetic significance of lichens have drawn attention to the findings of various researchers that certain elements such as phosphorus, calcium, magnesium, potassium, and iron are accumulated in lichen thalli and converted into forms which are available for the growth of subsequent colonizing plant species. An example given was that *Parmelia* spp. accumulate phosphorus while growing on maiaskites (a nephiline-bearing rock) and gneissic granites and that this element was present in the thallus in a 400-fold increase over that of the unweathered rock.[41] Phosphorus was enriched 90-fold in the fine-earth fraction beneath certain lichen thalli and existed as organic and Fe-bound phosphate. In the same review, the ability of lichens to accumulate potassium and iron was also noted.

Smith[42] has drawn attention to the fact that since the inorganic nutrient supply in the usually barren habitats of lichens is meager, the majority of lichen species possess a very efficient mechanism for nutrient accumulation.

For a comprehensive account of the absorption and accumulation of elements, the reader is referred to Tuominen and Jaakola.[43] They drew attention to the fact that lichens differ considerably from higher plants and mosses in trace element concentrations and commented that most of the elements studied by Lounamaa[44] occurred in greater amounts in lichens than in other plants, with manganese being the exception, occurring in lower amounts in lichens. Cobalt, nickel, molybdenum, and silver occurred in the same amounts in lichens as in other plants from corresponding habits.

In a study on the trace elements on some Snowdonian rocks, their minerals and related soils, Jenkins[45] concluded that both atmospheric sources and substratum contributed to the trace element contents of various saxicolous lichens, including *Lecanora gangaloides, Haematomma ventosum, Rhizocarpon geographicum* (all crustose species), *Parmelia omphalodes*, and *P. saxatilis* (foliose species).

Because the subsequent studies on the effect of lichen acids, extracted from *P. omphalodes*, on ground basic rock samples were inconclusive, Jenkins[45] suggested that individual lichen acids should have been allowed to react with defined rock minerals at low temperature. Interestingly, from thin section observations, he showed that hyphae of *P. omphalodes* penetrated rock as was the case with *R. geographicum* on a rock substrate. Jenkins[45] found no evidence of chemical alteration of individual minerals in sections of rocks (plagioclase felspar) colonized by *P. omphalodes*, but this was probably largely due to the fact that the optical microscope available to the author was unable to detect signs of weathering which the scanning electron microscope would have revealed. The author finally discussed the significance of lichens in pedogenesis in the light of his findings and concluded that they can be an important factor in the initial stages, although geologically they are ephemeral.

C. Soil Development in Different Environments

An interesting study by Krumbein and Jens[46] discussed the biogenic "rock varnishes" of the Negev Desert in Israel. The varnishes have been designated as "weathering crusts", "desert rinds", and "protective coatings", and these authors considered the term "rock varnish" should replace the misleading term "desert varnish". This varnish consists mainly of inorganic components such as iron and manganese oxides with some quartz, clays, and carbonates admixed, but in addition may contain considerable amounts of organic material giving rise to the gel-like or lacquer-like appearance. Where these coatings consist mainly of cyanobacteria, the manganese and iron contents are low. These elements may come from a variety of sources including the rock, running water, dust, rain, and the surrounding soil. In their thorough search of the literature, Krumbein and Jens[46] noted that the possibility of algae and lichens being implicated in this phenomenon had been suggested by various researchers.[47,48]

In their paper, Krumbein and Jens[46] referred to perforation patterns of the rock produced by the fruiting bodies of endolithic lichens or by fungi and cyanobacteria. The pitting and exfoliation of the rock are referred to as "microbial solution and disintegration fronts",

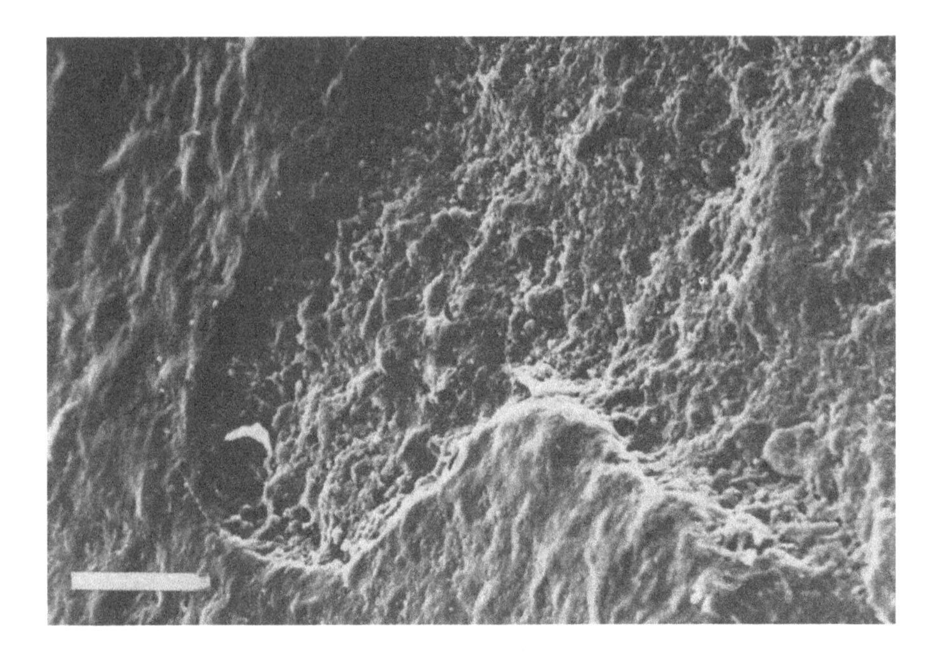

FIGURE 3. SEM micrograph of an endolithic lichen destroying a varnished chert. (Scale bar = 60 μm.) (Courtesy of Prof. W. Krumbein.)

which result in the iron and manganese coatings. A scanning electron micrograph illustrates an endolithic lichen which has destroyed (etched) a portion of varnished chert (Figure 3). Also shown is a detail of this zone illustrating a fruiting body of the endolithic lichen projecting from its rock channel with clay particles on its surface (Figure 4). The overall conclusion of these authors was that the rock varnish, mainly produced by the activity of lichenized epi- and endolithic cyanobacteria, chemoorganotropic bacteria, and fungi, plays a protective role for the underlying microflora which would otherwise be directly exposed to the harsh desert conditions.

In a study of desert varnish on exposed rock surfaces, Potter and Rossman[49] discussed the importance of clay minerals which comprised more than 70% of the specimen they examined from the Mojave Desert, California. In their article, they referred to an orange coat which develops in contact with soil on the bottom of the desert pavement stones. They characterized their specimens by IR spectroscopy, X-ray diffraction, and electron microscopy, and detected illite, illite-montmorillonite with small amounts of kaolinite, and sometimes chlorite. The orange coat gave a residue of approximately 10%, consisting mainly of iron. They concluded that the clay had been transported to the rock surface, but that clay may be an active agent in desert varnish formation. There were no microbiological assays made by these authors.

Laudermilk[48] has reviewed earlier work on desert varnish occurring in various parts of the world, including Africa, South America, the United States, and Europe. The author noted that Francis[47] considered a lichen to be contributing to the phenomenon in Australia and this agreed with the studies on desert varnish in Southern California. Interestingly, one rock specimen in section showed evidence of weathering in association with the varnish. The conclusion of the work was that the lichen secreted acids which corroded the outer layers of the rock surface and that part of the dissolved iron and manganese salts was precipitated on the surrounding rock as hydroxides.

During a study in Israel on patterns of limestone and dolomite weathering by lichens and algae, Danin et al.[50] drew attention to an intricate pattern, resembling a jigsaw puzzle, caused

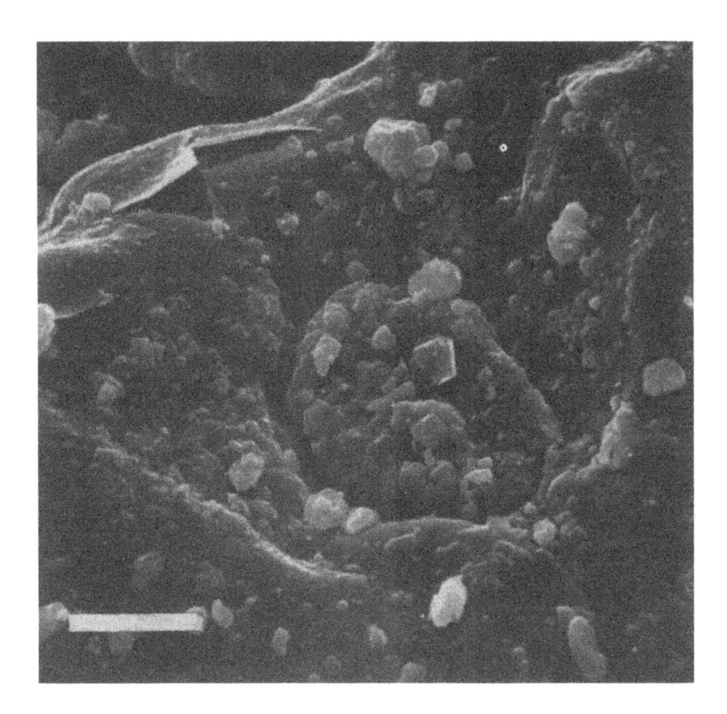

FIGURE 4. Enlargement from Figure 3 with a fruiting body of the endolithic lichen projecting from its rock channel. Clay particles can be seen on the varnish-coated rock as well as on the fruiting body. (Scale bar = 15 μm.) (Courtesy of Prof. W. Krumbein.)

by the colonization of the rock by lichens. In the Hizma area of Jerusalem, they found that the grey rocks are populated mainly with colonies of the endolithic lichen *Caloplaca alociza*. Microgrooves, 0.1 mm deep, occur between these colonies and are presumably the result of lichen activity. In contrast, on hillslopes near the mouth of Nahal Qidron, micropitting of massive limestome was the result of the cyanobacteria, *Gloeocapsa* spp.

A recent publication[51] on the weathering effects of crustose lichens on mica-schist rock from Signy Island in Antarctica described the considerable mechanical disruption of the rock surface.

In a study of the microbiology on rock surfaces and weathered stones under more temperate climatic conditions, Webley et al.[52] were able to show that an increasing number of microorganisms were associated with an increasing degree of colonization of rock surfaces by lichens. As expected, the highest numbers were generally found in raw soil in rock crevices where the organic matter content was high. In addition, bacteria, fungi, and actinomycetes were found only in the interior of porous weathered stones and not in unweathered stones. Of significance was the high proportion of microorganisms which were able to render silicates soluble in pure culture. The rock sequences selected for study were those which showed a succession in degree of colonization by lichens, mosses, and flowering plants. The predominant rock was acid igneous and metamorphic, but some were basic. Among the lichens encountered were the foliose species *Xanthoria parietina, Hypogymnia physodes*, and *Platysma glaucum* and the crustose lichen *Pertusaria sp*. Interestingly, the most active fungi in relation to the ability to render soluble silicates were those which produced citric acid and oxalic acid. Of the bacterial isolates, those producing 2-ketogluconic acid were the most active in dissolving silicates.

IV. CONCLUSIONS

There now seems little doubt that lichens, through their biophysical and biogeochemical activities, alter significantly the rock surfaces which they colonize. This must mean that disintegration of the rock into small particles, themselves corroded to produce a larger surface area, will contribute to the mineral matter of a developing soil. The production of oxalic acids and other lichen acids by lichen species colonizing soil surfaces must also contribute to pedological processes, leading to the formation of such compounds as oxalates which play an important part in chemical processes in soil. In addition, the organic fraction of soil is augmented by the accumulation of lichen residues after their death and transformation by degradative processes. Perhaps the most significant contribution of lichens to the formation of primitive soils is to be found in extreme environments such as the desert and arctic and antarctic conditions, referred to in this review.

Future studies on implications of lichen weathering of mineral substrates to the formation of primitive soils will depend on the application of modern techniques such as electron microscopy, coupled with microprobe analysis as a particularly important method, to the field. In addition, such studies would be meaningless without the collaborative efforts of mineralogists and microbiologists.

REFERENCES

1. **Sollas, W. J.,** On the activity of a lichen on limestone, *Rep. Br. Assoc. Adv. Sci.*, p. 586, 1880.
2. **Cooper, R. and Rudolph, E. D.,** The role of lichens in soil formation and plant succession, *Ecology*, 34, 805, 1953.
3. **Clements, F. E. and Shelford, V. E.,** *Bio-ecology*, Champman & Hall, London, 1939, 76.
4. **Schatz, A.,** The importance of metal-binding phenomena in the chemistry and microbiology of the soil. I. The chelating properties of lichens and lichen acids, *Adv. Front. Plant Sci.*, 6, 113, 1963.
5. **Silverman, M. P.,** Biological and organic chemical decomposition of silicates, in *Studies in Environmental Science 3. Biochemical Cycling of Mineral-Forming Elements*, Trudinger, P. A. and Swaine, D. J., Eds., Elsevier, New York, 1979, 445—465.
6. **Wilson, M. J. and Jones, D.,** The occurrence and significance of manganese oxalate in *Pertusaria corallina* (Lichenes), *Pedobiologia*, 26, 373, 1984.
7. **Jones, D., Wilson, M. J., and McHardy, W. J.,** Lichen weathering of rock-forming minerals: application of scanning electron microscopy and microprobe analysis, *J. Microsc.*, 124, 95, 1981.
8. **Fry, E. J.,** The mechanical action of crustaceous lichens on substrata of shale, schist, gneiss, limestone and obsidian, *Ann. Bot.*, 41, 437, 1927.
9. **Fry, E. J.,** A suggested explanation of the mechanical action of lithophytic lichens on rocks (shale), *Ann. Bot.*, 38, 175, 1924.
10. **Jacks, G. V.,** Organic weathering, *Sci. Prog. (London)*, 41, 301, 1953.
11. **Polynov, B. B.,** First stages of soil formation on massive crystalline rocks, *Pochvovedenie*, 7, 327, 1945.
12. **Bobriskaya, M. A.,** The absorption of mineral elements from massive crystalline rocks by lithophilous flora, *T. Pochv. Inst.*, 5, 1950.
13. **Schatz, A., Cheronis, N. D., Schatz, V., and Trelawny, G. S.,** Chelation (sequestration) as a biological weathering factor in pedogenesis, *Proc. Pa. Acad. Sci.*, 28, 44, 1954.
14. **Schatz, A.,** Soil microorganisms and soil chelation. The pedogenic action of lichens and lichen acids, *Agric. Food Chem.*, 11, 112, 1963.
15. **Hale, M. E., Jr.,** *Lichen Handbook*, Smithsonian Institute, Washington, D.C., 1961.
16. **Syers, J. K.,** Chelating ability of fumarprotocetraric acid and *Parmelia conspersa*, *Plant Soil*, 31, 205, 1969.
17. **Dawson, H. J., Hrutfiord, B. F., and Ugolini, F. C.,** Mobility of lichen compounds from *Cladonia mitis* in arctic soils, *Soil Sci.*, 138, 40, 1984.
18. **Iskandar, I. K. and Syers, J. K.,** Metal-complex formation by lichen compounds, *J. Soil Sci.*, 23, 255, 1972.

19. **Williams, M. E. and Rudolph, E. M.,** The role of lichens and associated fungi in the chemical weathering of rock, *Mycologia*, 66, 648, 1974.
20. **Schatz, V., Schatz, A., Trelawny, G. S., and Barth, K.,** Significance of lichens as pedogenic (soil forming) agents, *Proc. Pa. Acad. Sci.*, 30, 62, 1956.
21. **Ascaso, C., Galvan, J., and Ortega, C.,** The pedogenic action of *Parmelia conspersa, Rhizocarpon geographicum* and *Umbilicaria pustulata, Lichenologist*, 8, 151, 1976.
22. **Jones, D., Wilson, M. J., and Tait, J. M.,** Weathering of a basalt by *Pertusaria corallina, Lichenologist*, 12, 277, 1980.
23. **Wilson, M. J., Jones, D., and McHardy, W. J.,** The weathering of serpentinite by *Lecanora atra, Lichenologist*, 13, 167, 1981.
24. **Wilson, M. J. and Jones, D.,** Lichen weathering of minerals: implications for pedogenesis, in *Residual Deposits: Surface-Related Weathering Processes and Materials*, Wilson, R. C. L., Ed., Special Publ. No. 11 Geol. Soc., Blackwell Scientific, London, 1983, 5—12.
25. **Ascaso, C., Galvan, J., and Rodriguez-Pascual, C.,** The weathering of calcareous rocks by lichens, *Pedobiologia*, 24, 219, 1982.
26. **Vidrich, V., Cecconi, C. A., Ristori, G. G., and Fusi, Pi.,** Verwitterung Toskanischer Gesteine unter Mitwirkung von Flechten, *Z. Pflanz. Bodenk.*, 145, 384, 1982.
27. **Graustein, W. C., Cromack, K., and Sollins, P.,** Calcium oxalate: occurrence in soils and effect on nutrient and geochemical cycles, *Science*, 198, 1252, 1977.
28. **Cromack, K., Sollins, P., Graustein, W. C., Spiedel, K., Todd, A. W., Spycher, G., Lis, C. Y., and Todd, R. L.,** Calcium oxalate accumulation and soil weathering in mats of the hypogeous fungus *Hysterangium crassum, Soil Biol. Biochem.*, 11, 463, 1979.
29. **Malajczuk, N. and Cromack, K.,** Accumulation of calcium oxalate in the mantle of ectomycorrhizal roots of *Pinus radiata* and *Eucalyptus marginata, New Phytol.*, 92, 527, 1982.
30. **Hallbauer, D. K. and Jahns, H. M.,** Attack of lichens on quartzitic rock surfaces, *Lichenologist*, 9, 119, 1977.
31. **Pomar, L., Esteban, M., Llimona, X., and Fontarnau, R.,** Acción de liquenes, algas y hongos en la telodiagénesis de las rocas carbonatadas de la zona litoral prelitoral Catalana, *Inst. Invest. Geol. Univ. Barcelona*, 30, 83, 1975.
32. **Golubic, S., Friedmann, I., and Schneider, J.,** The lithobiontic ecological niche, with special reference to microorganisms, *J. Sediment. Petrol.*, 51, 475, 1981.
33. **Friedmann, E. I.,** Endolithic microorganisms in the Antarctic cold desert, *Science*, 215, 1045, 1982.
34. **Galvan, J., Rodriguez, C., and Ascaso, C.,** The pedogenic action of lichens in metamorphic rocks, *Pedobologia*, 21, 60, 1981.
35. **Ugolini, F. C. and Edmonds, R. L.,** Soil biology, in *Pedogenesis and Soil Taxonomy*, Wilding, L., Smeak, N., and Hall G., Eds., Elsevier, New York, 1983, 193—231.
36. **Oosting, H. J. and Anderson, L. E.,** The vegetation of a barefaced cliff in Western North Carolina, *Ecology*, 18, 280, 1937.
37. **Keever, C., Oosting, H. J., and Anderson, L. E.,** Plant succession on exposed granite of rocky face mountain, Alexander county, North Carolina, *Bull. Torrey Bot. Club*, 78, 401, 1951.
38. **Krusentjerna, E.,** The growth on rock, *Acta Phytogeogr.*, 50, 144, 1965.
39. **Yarilova, E. A.,** The role of lithophilic lichens in weathering crystalline bed rock, *Pochvovedenie*, 42, 533, 1947.
40. **Syers, J. K. and Iskandar, I. K.,** Pedogenetic significance of lichens, in *The Lichens*, Ahmadjian, V. and Hale, M. E., Eds., Academic Press, New York, 1973, 225—248.
41. **Lazarev, A. A.,** The accumulation and transformation of phosphorous on miaskites and granite-gneisses in the earliest stages of soil formation, *Pochvovedenie*, 7, 340, 1945.
42. **Smith, D. C.,** The biology of lichen thalli, *Biol. Rev.*, 37, 537, 1962.
43. **Tuominen, Y. and Jaakola, T.,** Absorption and accumulation of mineral elements and radioactive nuclides, in *The Lichens*, Ahmadjian, V. and Hale, M. E., Eds., Academic Press, New York, 1973, 185—223.
44. **Lounamaa, J.,** Trace elements in plants growing wild on different rocks in Finland, *Ann. Bot. Soc. Zool. Bot. Fenn. Vanamo*, 29, 1, 1956.
45. **Jenkins, D. A.,** Trace Element Studies on Some Snowdonian Rocks, Their Minerals and Related Soils, Ph.D. thesis, University of Bangor, Wales, 1964.
46. **Krumbein, W. E. and Jens, K.,** Biogenic rock varnishes of the Negev Desert (Israel); an ecological study of iron and manganese transformation by cyanobacteria and fungi, *Oecologia*, 50, 25, 1981.
47. **Francis, W. D.,** The origin of black coatings of iron and manganese oxides on rocks, *Proc. R. Soc. Queensl.*, 32, 110, 1920.
48. **Laudermilk, J. D.,** On the origin of desert varnish, *Am. J. Sci.*, 21, 51, 1931.
49. **Potter, R. M. and Rossman, G. R.,** Desert varnish: the importance of clay minerals, *Science*, 196, 1446, 1977.

50. **Danin, A., Gerson, R., Marton, K., and Garty, J.,** Patterns of limestone and dolomite weathering by lichens and blue-green algae and their palaeoclimatic significance, *Palaeogeogr. Palaeoclimatol. Palaeoecol.*, 37, 221, 1982.
51. **Walton, D. W. H.,** A preliminary study of the action of crustose lichens on rock surfaces in Antarctica, in *Antarctic Nutrient Cycles and Food Webs*, Siegfried, W. R., Condy, P. R., and Laws, R. M., Eds., Springer-Verlag, Berlin, 1985, 180.
52. **Webley, D. M., Henderson, M. E. K., and Taylor, I. F.,** The microbiology of rocks and weathered stones, *J. Soil Sci.*, 14, 102, 1963.

Section XIII: Methods for Cultivating Lichens and Isolated Bionts

Chapter XIII

METHODS FOR CULTIVATING LICHENS AND ISOLATED BIONTS

Paul Bubrick

I. THALLUS

For most experimental work, field-collected lichens should be used. Depending upon the type, lichens may be stored in the laboratory in a desiccated state or by freezing.[1-3] However, stored lichens should be routinely compared to freshly collected lichens to ensure that the studied parameters do not deteriorate over time. Prior to experiments, thalli are cleaned and allowed a "recovery period" to come to equilibrium with their surroundings. Such a recovery period can be very important since rewetting of dry thalli may cause dramatic physiological effects.[4] Subsequent studies may be done on whole thalli,[1,5] thalli parts,[6] punched disks,[7] or excised tissues (e.g., cephalodia[8]). A discussion on the merits of flow systems vs. closed-loop and dish systems for the study of whole lichen physiology has been presented by Kershaw.[4]

Certain lichens can be maintained or grown under laboratory conditions. Growth chambers may be simple[9-12] or complex,[13-15] including walk-in chambers[2] or phytotrons.[16] Substrates for lichen growth include soil, sand, filter paper,[17] and sponge plastic.[10] Silica gel has proved effective for the growth of a number of cyanolichens,[18] or the maintenance of certain lichens with eukaryotic phycobionts.[11] When pollution-sensitive lichens are studied under city conditions, special air filtration devices may be required.[13,15]

The environmental conditions necessary for the maintenance of intact lichens have been discussed by Ahmadjian.[17] One important factor is the alternate wetting and slow drying of thalli. In addition, nutrients supplied during inoculation, or during subsequent irrigation, should be minimal. Overirrigation or excess nutrients usually leads to the death or dissolution of the symbiosis. Other environmental factors (e.g., light and temperature) must be optimized for individual lichens.[9-15]

Cultures of lichens can be produced in the laboratory by resynthesis of the isolated symbionts. The two most successful protocols have been developed by Ahmadjian et al.[19,20] These involve the mixing of isolated bionts and inoculation onto soil[19] or newly cleaved mica[20] substrates. Soil in clay pots or mica in flasks are incubated at 2000 to 3000 lux and 18 to 21°C under 12 hr light per 12 hr dark regimes. Once resynthesized, thalli may be grown under phytotron conditions.[21]

Resynthesized thalli develop only after several weeks or months.

II. PHOTOBIONT

Examination of lichens in a dissecting microscope often provides valuable clues for the isolation of photobionts. In some lichens (e.g., *Peltigera*), the fungal medulla can be peeled away with forceps leaving tissue highly enriched with photobiont cells. In other lichens (e.g., *Dermatocarpon*), photobiont cells are near the thallus surface and can be scraped off with a knife. Visual inspection may also reveal epiphytes and debris which can be removed with forceps or a soft-bristled brush.

Usually, thalli must be disrupted before photobionts can be isolated. Disruption is best done gently by hand in a mortar as fragile cells may be broken in blenders. When disrupting thalli in buffers, it may be helpful to omit calcium ions (or replace them with magnesium ions) since Ca^{2+} may cause flocculation of photobiont cells and hyphae.[66]

For some experiments, large quantities of highly purified, freshly isolated photobionts are required. A number of different types of protocols have been developed, including differential centrifugation, gradient centrifugation, and gel filtration.

Protocols for the purification of photobionts by differential centrifugation have been published by several workers.[22-26] There are no standard protocols; optimal conditions for each algal type must be empirically determined. For small photobionts (e.g., *Coccomyxa*), low-speed centrifugation (e.g., less than 500 rpm) in a clinical centrifuge may leave the bulk of the algal cells in the supernatant. For larger photobionts (e.g., *Trebouxia*), various time-speed combinations must be tested. Low-speed centrifugation ($100 \times g$) is used to remove large fragments of thalli or clumps of tissue; higher speeds ($400 \times g$) are used to separate intact cells from cell fragments, debris, etc. This type of work should be done near a light microscope so that the progress of the purification can be monitored. Such procotols usually take 30 to 60 min and result in 20 to 70% recovery.

Prior to differential centrifugation, disrupted thalli may be filtered through a plastic strainer,[23] sequentially through teflon screens[27] (e.g., 30, 20, and 10 μm diameter pore size), or through screens or large-pored filters by vacuum filtration.

Gradient centrifugation has been used to obtain highly purified, freshly isolated photobionts. Ascasco[28] described the use of $CsCl_2$ step gradients for isolation of *Trebouxia*. Following disruption and washings, cells were suspended in sucrose (0.25 M, 2 mℓ) and gently layered onto a solution of $CsCl_2$ (density 1.550 g cm $-{}^{3}$, 3 mℓ). Tubes were centrifuged in a swinging bucket rotor for 10 min at 4500 rpm (g not given). Purified cells were recovered from the sucrose-$CsCl_2$ interface and extensively washed. Step gradients using KI (80% w/v) may also be employed, although KI may be harmful to cells after prolonged exposure.[29] Percoll® (Pharmacia, Uppsala, Sweden) has been used to purify *Pseudotrebouxia*.[29] Percoll® was first diluted according to the manufacturer's instructions. Photobionts were added to Percoll®, and self-generated, continuous gradients were formed at $30{,}000 \times g$ for 15 to 20 min in a fixed-angle centrifuge. By using density marker beads (Pharmacia), photobiont cell density can be determined and conditions optimized. A comparison between $CsCl_2$, KI, and Percoll® gradients for the purification of the *Ramalina duriaei* phycobiont *(Pseudotrebouxia)* is illustrated in Figure 1.

A procedure for the isolation and purification of *Coccomyxa* using discontinuous sucrose gradients has been described by Honegger and Brunner.[30]

Protocols involving gradient centrifugation can be completed in 15 to 30 min and result in up to 60% recovery.

Recently, the purification of the *Evernia prunastri* phycobiont by gel filtration has been described.[31] Thalli were first floated on buffer, then macerated, filtered, and taken through a short cycle of differential centrifugation. Algae were dispersed and loaded onto a column of Sepharose 2B (Pharmacia) (height 2 cm, diameter 3.5 cm) equilibrated with distilled water; algae were eluted with distilled water. Purity was monitored both visually and by quantitation of polyols (ribotol vs. mannitol); recovery of cells was 30%.

In some studies, the final aim may be the isolation, purification, and growth of photobionts in culture. One suitable method for the isolation and purification of photobionts is with the aid of micropipettes.[17,26,32] Micropipettes are pulled from sterilized, 3 to 4-mm glass tubes (e.g., Pasteur pipettes) or from capillary tubes. The tip of the pipette is grasped with forceps and the region behind the tip is heated in a flame until it softens. In one action, the tube is removed from the flame and the glass stretched with a rapid smooth pull. The tip is checked for an opening (20 to 70 μm); if no opening is present, the tip is broken off with fine forceps. Lichen thalli are then disrupted by griding between two glass surfaces. A drop of the suspension is placed on a slide and examined in the light microscope. Photobiont cells with hyphal fragments attached to their walls are transferred with the micropipette through successive sterile water drops on the slide until visible impurities are removed. Transferring

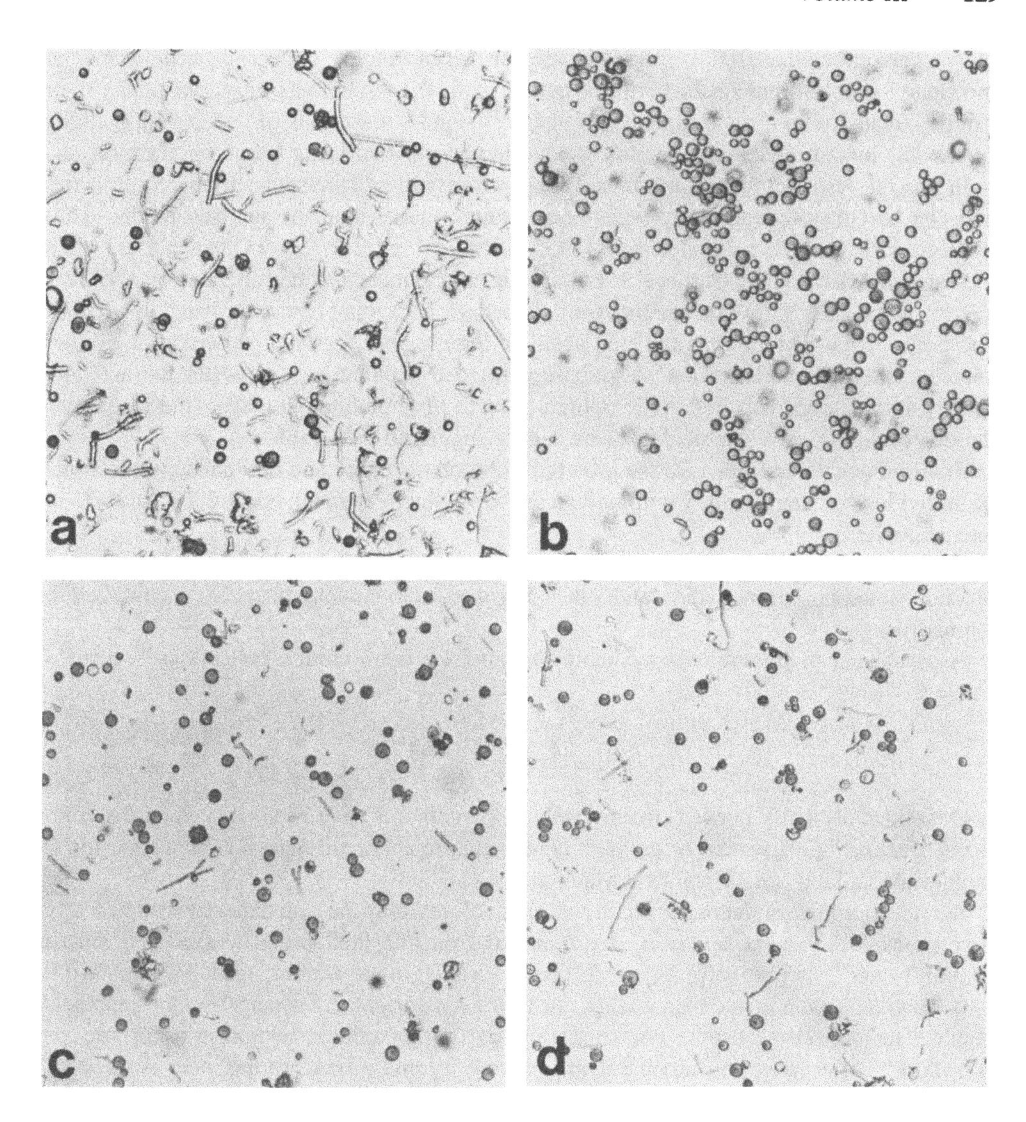

FIGURE 1. Freshly isolated and purified phycobionts from *Ramalina duriaei*. Phycobionts purified by (a) differential centrifugation, (b) $CsCl_2$ step gradients, (c) KI step gradients, and (d) Percoll®.

can be done by attaching rubber tubing to the thick end of the micropipette and using gentle suction. After transferring through water drops, cells are inoculated into suitable media. Although this method may be somewhat time consuming and difficult to master, it can result in axenic and clonal cultures of photobionts in one step.

Micropipettes may be less suitable for photobionts without attached hyphae or for very small photobionts. In such cases, photobionts may be partially purified by differential or gradient centrifugation. At this stage, most of the techniques used for the purification of microalgae and cyanobacteria can be applied to photobionts. A synopsis of useful techniques can be found in Stein.[33] Thus, photobionts may be streaked on an agar surface and purified cells or colonies picked. For nitrogen-fixing cyanobionts, nitrogen-free media may be employed. Cyanobionts which produce hormogonia (e.g., *Nostoc*) may be purified by taking advantage of the fact that motile hormogonia purify themselves as they move across the surface of the agar.

Photobionts may also be isolated as they grow from thallus fragments incubated in liquid or on agarized media. Thallus fragments are washed in running water and sterile water, and inoculated into nutrient media. After a period of time, photobionts grow out of thalli and can be isolated with a micropipette or forceps.[34] Hand-cut sections of lichen thalli can also be used as inocula.[35] Unialgal cultures may then be purified using routine procedures.

In general, purified photobionts do not appear to have highly unusual nutrient requirements, and can be cultivated on many types of commonly used liquid or agarized media. Thus, cyanobacteria may be cultivated on KM*,[36] ASM-1*,[37] or BG-11*.[38] Other nutrient media are listed in Allen[39,40] and Carr et al.[41] Eukaryotic phycobionts may be cultivated on *Trebouxia* agar*[26,32] and BBM*[42] with or without supplements*; additional media are listed in Nichols[43] and Starr.[44] Few photobionts have an absolute requirement for organic carbon or nitrogen sources, although the addition of glucose, proteose peptone, or soil-water extract* may greatly stimulate growth.[17,26,32] For optimal growth of photobionts on agar, the recommendations of Allen[39] and Archibald[45,46] are highly suggested (see Table 1.)

Environmental conditions for the growth of photobionts vary and optimal conditions must be individually determined. Photobionts can be routinely maintained at 2000 to 3000 lux and 15 to 22°C.[45-47]

An extensive list of organic carbon and nitrogen sources for photobionts, including sugar, alcohol, vitamin, amino acid, and other nitrogen compounds, have been compiled by Ahmadjian.[48]

A partial list of photobionts available for purchase from culture collections is given in Table 2.

III. MYCOBIONT

Quantities of highly purifed, freshly isolated mycobionts have been isolated by dissection of the thallus.[49] Differential or gradient centrifugation apparently has not been used for the large-scale isolation of symbiotic mycobionts.

For the purpose of laboratory culture, mycobionts may be conveniently isolated from ascospores.[17,26] Peri- or apothecia are removed from the thallus and washed (60 min) in running water. Alternatively, ascocarps may be soaked in the wetting agent Tween 80 (0.01 to 0.1% v/v) for 30 min and then washed. In *Xanthoria parietina*, Tween 80-washed apothecia resulted in reduced ascospore contamination and no apparent reduction in ascospore germination.[66] After washing, ascocarps are blotted dry and affixed to the inner surface of a petri dish lid with a small daub of vaseline. As ascocarps dry, asci open and spores are ejected onto an agar surface. After ejection (hours to days), lids are removed and replaced with clean lids. Plates are sealed and incubated at 10 to 25°C until ascospores germinate (days to months). Agar blocks containing germinated ascospores free of contamination are cut out and transferred to suitable liquid or agarized medium.

Ascospores can be germinated on nutrient-poor media (agarized 1 to 2% w/v tap water is sufficient); high nutrient content may inhibit the germination of many types of ascospores.[50] The effects of various environmental conditions on ascospore germination have been summarized by Pyatt.[50]

In some lichens, ascospores either do not germinate on agar or cease growth immediately after germination. In such cases, germination was stimulated by including photobiont extracts in the medium;[51] growth was stimulated by adding whole photobiont cells to the agar near the fungal ascospores.[52]

For short-term experiments, cultivation of mycobionts can be conveniently done on microscope slides.[53,54] Slides may be dipped into media, or 1 to 3 mℓ of media pipetted onto the slide. Ascospores are then discharged and germinated on the slides as described.

* Compositions are given in Table 1.

Table 1
COMPOSITION OF NUTRIENT MEDIA

1. Photobionts

A. ASM-1[37]

Nutrient stock solutions (g/100 mℓ distilled water)

1. $NaNO_3$	1.698	5. $MgSO_4 \cdot 7H_2O$	0.492
2. $MgCl_2 \cdot 6H_2O$	0.406	6. $CaCl_2$	0.222
3. K_2HPO_4	0.174	7. $FeCl_3 \cdot 6H_2O$	0.010
4. Na_2HPO_4	0.268	8. $Na_2 \cdot EDTA$	0.058

Trace element stock solution

1. To 100 mℓ of acidified water (99.9 mℓ distilled water + 0.1 mℓ concentrated H_2SO_4), add $MnCl \cdot 4H_2O$ (0.014 g), $CoCl_2 \cdot 6H_2$) (2 × 10 g), H_3BO_3 (0.024 g), $ZnCl_2$ (0.004 g), $CuCl_2 \cdot 2H_2O$(1 × 10^{-6} g).

To 919 mℓ of distilled water, add 10 mℓ of each nutrient stock solution and 1 mℓ of the trace element stock solution. After autoclaving and cooling, pH 6.8 to 7.0. Medium may be agarized (1 to 1.5% w/v).

B. BBM[42]

Nutrient stock solutions (g/100 mℓ distilled water)

1. $NaNO_3$	2.5	4. K_2HPO_4	0.75
2. $MgSO_4 \cdot 7H_2O$	0.75	5. KH_2PO_4	1.75
3. $CaCl_2$	0.25	6. NaCl	0.25

Trace element stock solutions

1. To 100 mℓ distilled water, add $Na_2 \cdot EDTA$ (5.0 g) and KOH (85%) (3.1 g).
2. To 100 mℓ distilled water, add H_3BO_3 (1.142 g).
3. To 100 mℓ acidified water (99.9 mℓ distilled water + 0.1 mℓ concentrated H_2SO_4), add $FeSO_4 \cdot 7H_2O$ (0.498 g).
4. To 100 mℓ acidified water, add $ZnSO_4$ (0.882 g), $Na_2MoO_4 \cdot 2H_2O$ (0.17 g), $Co(NO_3)_2 \cdot 6H_2O$ (0.049 g), $MnCl_2$ (0.144 g), $CuSO_4 \cdot 5H_2O$ (0.157 g).

To 936 mℓ of distilled water, add 10 mℓ of each nutrient stock solution and 1 mℓ of each trace element stock solution. After autoclaving and cooling, pH 6.6. Medium may be agarized (1 to 1.5% w/v).

Common supplements include 3NBBM[46] (all stock solutions as above except $NaNO_3$, 7.5 g/100 mℓ), glucose (10 g/ℓ), proteose peptone (10 g/ℓ, or soil-water extract 10 mℓ/ℓ) (see I.E).

C. BG-11[38]

Nutrient stock solutions (g/100 mℓ distilled water)

1. $NaNO_3$	15.0	5. Citric acid	0.06
2. K_2HPO_4	0.4	6. Fe ammonium citrate	0.06
3. $MgSO_4 \cdot 7H_2O$	0.75	7. $Na_2 \cdot EDTA$	0.01
4. $CaCl_2 \cdot 2H_2O$	0.36	8. Na_2CO_3	0.2

Trace element stock solution

1. To 100 mℓ distilled water, add H_3BO_3 (0.286 g), $MnCl_2 \cdot 4H_2O$ (0.181 g), $ZnSO_4 \cdot 7H_2O$ (0.0222

To 919 mℓ distilled water, add 10 mℓ of each nutrient stock solution and 1 mℓ of the trace element stock solution. After autoclaving and cooling, pH 7.1. Medium may be agarized (1 to 1.5% w/v).

D. KM[36]

Nutrient stock solutions (g/100 mℓ distilled water)

1. $Na_2 \cdot EDTA$	1.5	4. $NaNO_3$	10.0
KOH	0.31		
2. $Fe(SO_4)_3 \cdot 6H_2O$	0.04	5. $Ca(NO_3)_2 \cdot 4H_2O$	0.1
3. $MgSO_4 \cdot 7H_2O$	1.5	6. K_2HPO_4	10.0

Trace element stock solution

1. To 100 mℓ distilled water, add $ZnSO_4 \cdot 7H_2O$ (0.88 g), $MnCl_2 \cdot 4H_2O$ (0.14 g), MoO_3 (0.071 g), $CuSO_4 \cdot 5H_2O$ (0.157 g), $Co(NO_3)_2 \cdot 6H_2O$ (0.049 g).

To each 939 mℓ distilled water, add 10 mℓ of each nutrient stock solution and 1 mℓ of the trace element stock solution. After autoclaving and cooling, pH 7.5 Medium may be agarized. MoO_3 may be replaced with $Na_2MoO_4 \cdot 2H_2O$ (0.119 g).

E. Soil-water extract

Mix 1 vol garden loam with 2 vol distilled water and autoclave (few minutes) or steam for 1 hr. After soil particles have settled, the supernatant is removed and stored at 4°C or in small aliquots at −20°C.

Table 1 (continued)
COMPOSITION OF NUTRIENT MEDIA

F. *Trebouxia* medium[26,32]
To 970 mℓ BBM (see I.B), add proteose peptone (10 g) and glucose (20 g). Medium may be agarized (1 to 1.5% w/v).

II. Mycobionts

A. Bianchi[56]
To 1000 mℓ distilled water, add NH_4 tartarate (5.0 g), NH_4NO_3 (1.0 g), KH_2PO_4 (1.0 g), $MgSO_4 \cdot 7H_2O$ (0.5 g), NaCl (0.1 g), $CaCl_2 \cdot 2H_2O$ (0.1 g), sucrose (10.0 g), biotin (1×10^{-5} g), thiamin (0.005 g). Medium may be agarized (1 to 2% w/v).

B. Lilly and Barnett[57]
To 1000 mℓ distilled water, add glucose (10.0 g), asparagine (2.0 g), KH_2PO_4 (1.0 g), $MgSO_4 \cdot 7H_2O$ (0.5 g), $Fe(NO_3)_3 \cdot 9H_2O$ (2×10^{-4} g), $ZnSO_4 \cdot 7H_2O$ (2×10^{-4} g), $MnSO_4 \cdot 4H_2O$ (1×10^{-4} g), thiamin (1×10^{-4} g), biotin (5×10^{-6} g). Medium may be agarized (1 to 2% w/v).

C. Malt-yeast extract[26,32]
To 1000 mℓ distilled water, add malt extract (20.0 g), yeast extract (2.0 g). Medium may be agarized (1 to 2% w/v).

D. Soil-extract medium[26,32]
To 960 mℓ BBM (see I.B), add soil-water extract (40 mℓ). Medium may be agarized (1 to 2% w/v).

Mycobionts have also been isolated from conidia.[55] Thalli are first washed, which causes the mucilage inside pycnidia to swell and the conidia to be extruded from the pycnidia ostiole. The conidial mass may be picked up with a needle or micropipette and inoculated onto suitable media.

In lichens lacking ascocarps, the mycobiont must be isolated from the thallus.[24,26,32] Mycobionts can be isolated from disrupted thalli with a micropipette. Mycobionts may also be isolated from dissected thalli or from thallus outgrowths (e.g., rhizines).

Suitable growth media for cultivation of mycobionts may be problematic since mycobionts appear to be somewhat individualistic in their nutrient requirements. Recommended media include Bianchi*,[56] malt-yeast extract*, soil-water medium*,[26,32] and Lilly and Barnett*[57] medium. Richardson[24] states that most mycobionts have a requirement for biotin and thiamin. In addition, shaking liquid cultures usually produces greater biomass than stationary liquid or agarized media.

An extensive list of organic carbon and nitrogen sources for mycobionts including carbon, amino acid, vitamin, and other nitrogen compounds has been compiled by Ahmadjian.[48]

A partial list of mycobionts available for purchase from culture collections is given in Table 3.

* Compositions are given in Table 1.

Table 2
PHOTOBIONTS AVAILABLE FROM CULTURE COLLECTIONS[a,b]

Asterochloris phycobiontica Tsch.-Woess (from *Varicellaria carneonivea)*; SAG B26.81
Chlorella saccharophila (Krueger) Nadson (from *Trapelia coarctata*); SAG 3.80
C. saccharophila (Krueger) Nadson (no lichens mentioned); CCAP 211/47, CCAP 221/48, CCAP 211/49, CCAP 211/50
Chroococcidiopsis sp.[c] (from *Anema mummularium)*, PU 1979/26322aPH
Chroococcidiopsis sp. (from *Peccania cerebriformis*); PU 1980/26370aPH
Chroococcidiopsis sp. (from *Psorotichia columnaris*); SAG B33.84, PU 1980/26391aPH
Chroococcidiopsis sp. (from *Gonohymenia* sp.); PU 1982/27283PH
"*Chroococcidiopsis-Myxosarcina*" (from *Peccania coralloides*); PU 1981/27690aPH
Coccobotrys verrucariae Chodat em. Vischer (no lichen mentioned); IBSG BS175
Coccomyxa mucigena Jaag (from *Peltigera aphthosa*); SAG 216-4, CCAP 216/4, UTEX 269
C. peltigerae Wáren (from *Peltigera aphthosa*); SAG 216-5, CCAP 216/5, UTEX 270
C. peltigerae Wáren var. *varilosae* Jaag (from *Peltigera variolosa*); SAG 216-6, CCAP 216/6, UTEX 271
C. pringsheimii Jaag (from *Botrydina vulgaris*); SAG 216-7, CCAP 216/7
C. pringsheimii Jaag (from *Botrydina vulgaris*); SAG 69.80
C. solorinae var. *bisporae* Jaag (from *Solorina bispora*); SAG 216-10, CCAP 216/10, UTEX 275
C. solorinae var. *croceae* Chodat (from *Solorina crocea*); SAG 216-11a, CCAP 216/11a, UTEX 276
C. solorinae var. *croceae* Chodat (from *Solorina crocea*); SAG 216-11b, CCAP 216/11b
C. solorinae var. *saccatae* Chodat (from *Solorina saccata*); SAG 216-12, CCAP 216/12, UTEX 277
C. subellipsoidea Acton (from *Botrydina vulgaris*?); SAG 216-13, CCAP 216/13
C. subellipsoidea Acton (from *Botrydina*); CCAP 216/15
Dictyochloropsis splendida Geitler (from *Chaenotheca brunneola*); SAG B244.80, CCAP 225/1
D. symbiontica Tsch.-Woess (from *Chaenothecopsis consociata*); SAG B27.81
Dilabifilum arthopyreniae (Vischer et Klemet) Tsch.-Woess (from *Verrucaria adriatica*); SAG 467-2, CCAP 415/2
D. incrustans (Vischer) Tsch.-Woess (from *Verrucaria aquatilis*); CCAP 415/1
Elliptochloris bilobata Tsch.-Woess (from *Catolechia wahlenbergii*); SAG 245.80
Entophysalidaceae (from *Gonohymenia nigritella*); PU 1981/27677PH
Hyalococcus dermatocarponis Wáren (from *Dermatocarpon fluviatilis*); UTEX 908
Leptosira obovata Vischer (from *Vezdaea aestivalis*); SAG B133.80
Mymecia biatorellae (Tsch.-Woess et Plessl) Petersen (from *Lobaria linita*); SAG B8.82
M. biatorellae (Tsch.-Woess et Plessl) Petersen (from *Dermatocarpon tuckermani*); UTEX 907
M. pyriformis Tsch.-Woess et Plessl (lichen unsure); CCAP 250/1
M. reticulata Tsch.-Woess (lichen unsure); CCAP 250/3
Myrmecia sp. (from *Lecidea crystallifera*); CCLS[d]
Nannochloris normandinae Tsch.-Woess (from *Normandina pulchella*); SAG B9.82
Nostoc sp. (from *Peltigera canina, P. polydactyla*); CCLS[d]
Prototheca wickerhamii Tubaki et Soneda[e] (from *Xanthoria parietina*); SAG 263-11
P. zopfii Kruger[e] (from *Xanthoria*); SAG 263-6
P. zopfii Kruger[e] (from *Xanthoria*, different collection); SAG 263-10
Prototheca sp.[e] (*Xanthoria parietina*); SAG 263-9
Pseudochlorella sp. (from *Lecidea granulosa, Stereocaulon strictum*); CCLS[d]
Pseudopleurococcus arthopyreniae Vischer (*Arthopyrenia kelpii*); IBSG BS654
Pseudotrebouxia aggregata Arch. (from *Xanthoria parietina*); SAG 219-1d, CCAP 219/1d, UTEX 180, IBSG IB235[f]
P. aggregata Arch. (from *Lecidea fuscoatra*); CCLS[d]
P. corticola Arch. (free-living); UTEX 909, IBSG IB326[f]
P. decolorans (Ahmad.) Arch. (from *Xanthoria parietina*); CCAP 219/4, UTEX 901, IBSG IB327[f]
P. decolorans (Ahmad.) Arch. (no lichen mentioned); CCAP 219/5a
P. decolorans (Ahmad.) Arch. (from *Buellia punctata*); UTEX 781
P. galapagensis Hildr. et Ahmad. (from *Ramalina* sp.); UTEX 2230, CCLS[g], IBSG IB333[f]
P. gigantea Hildr. et Ahmad. (from *Caloplaca cerina*); UTEX 2231, CCLS[g], IBSG IB334[f]
P. higginsiae Hildr. et Ahmad. (from *Buellia straminea*); UTEX 2232, CCLS[g], IBSG IB335[f]
P. impressa (Ahmad.) Arch. (from *Physcia stellaris*); UTEX 892, UTEX 893, IBSG 330, IBSG 331[f], CCLS[g]
P. incrustata (Ahmad.) Arch. (from *Lecanora dispersa*); UTEX 784, CCAP 219/6, IBSG IB329[f]
P. jamesii Hildr. et Ahmad. (from *Schaereria tenebrosa*); UTEX 2233, IBSG IB336,[f] CCLS[g]
P. potteri (Ahmad.) Arch. (from *Rhizoplaca chrysoleuca*); UTEX 900, IBSG IB332[f], CCAP 219/7
P. potteri (Ahmad.) Arch. (from *Pertusaria* sp.); CCLS[d]
P. showmanii Hildr. et Ahmad. (from *Lecanora hageni*); UTEX 2234, IBSG IB337[f]

Table 2
PHOTOBIONTS AVAILABLE FROM CULTURE COLLECTIONS[a,b]

P. usneae Hildr. et Ahmad. (from *Usnea filipendula*); UTEX 2235, IBSB IB338[f], CCLS[g]
Pseudotrebouxia sp. (from *Acarospora fuscata, Aspicilia calcarea, Astroplaca* sp., *Caloplaca holocarpa, Diploschistes scruposus, Lecania* sp., *Lecidea sarcogynoides, L. tenebrosa, Omphalodium arizonicum, Parmelia tinctorum, Rhizocarpon geographicum, Toninia caerulonigricans*); CCLS[d]
Trebouxia anticipata (Ahmad.) Arch. (from *Parmelia rudecta*); CCAP 219/3, IBSG IB340, UTEX 903
T. anticipata (Ahmad.) Arch. (from *Parmelia rudecta*); UTEX 904, IBSG IB341
T. arboricola de Puymaly (no lichen mentioned); SAG 219-1a, CCAP 219/1a, IBSG IB363
T. crenulata Arch. (from *Xanthoria aureola*); CCAP 219/2, IBSG IB359
T. crenulata Arch. (from *Parmelia acetabulum*) CCAP 219/1b
T. erici Ahmad. (from *Cladonia cristatella*); UTEX 910, UTEX 911, UTEX 912, IBSG IB342, IBSG IB343, IBSG IB344
T. erici Ahmad. (from *Cladonia subulata*); IBSG IB365
T. erici Ahmad. (from *Cladonia grayii*); IBSG IB364
T. excentrica Arch. (from *Sterocaulon dactyophyllum* var. *occidentale*); UTEX, 1714, IBSG IB345
T. excentrica Arch. (from *Cladonia bacillaris, C. subtenuis, C. leporina, Huilia tuberculosa, Lecidea metzleri, Lepraria* sp.); CCLS[d]
T. flava Arch. (from *Physcia pulverulenta*); CCAP 219/1c, UTEX 181, IBSG IB346
T. gelatinosa (Ahmad.) Arch. (from *Parmelia caperata*); UTEX 905, UTEX 906, IBSG IB347, IBSG IB348
T. gelatinosa (Ahmad.) Arch. (from *Hypogymnia physodes*); IBSG IB366
T. glomerata (Wáren) Ahmad. (from *Stereocaulon evolutoides*); UTEX 894, UTEX 895, IBSG IB349, IBSG IB350
T. glomerata (Wáren) Ahmad. (from *Stereocaulon pileatum*); UTEX 896, UTEX 897, IBSG IB351, IBSG IB352
T. glomerata (Wáren) Ahmad. (from *Cladonia boryi, Huilia albocaerulescens, Stereocaulon saxatile*); CCLS[d]
T. irregularis Hildr. et Ahmad. (from *Stereocaulon* sp.); UTEX 2236, IBSG IB339, CCLS[g]
T. italiana Arch. (from *Xanthoria parietina*); CCAP 219/5b, IBSG IB358
T. magma Arch. (from *Pilophorus acicularis*); UTEX 902, IBSG IB354
T. magma Arch. (*Cladonia* sp.); ATCC 30426, SAG 100.80, CCAP 213/3, UTEX 67
T. pyriformis Arch. (from *Cladonia squamosa*); UTEX 1712, IBSG IB355
T. pyriformis Arch. (from *Stereocaulon pileatum*); UTEX 1713, IBSG IB356, CCLS[g]
T. simplex Tsch.-Woess (from *Chaenotheca chrysocephala*); SAG B101.80, IBSG IB361
Trebouxia sp. (from *Cladonia elongata*); IBSG V235[h]
Trebouxia sp. (from *Cladonia chlorophaea, C. verticillata, Gymnoderma lineare, Lecidea crustulata, L. erratica, Lepraria zonata, Parmeliopsis hyperopta, Physcia millegrana, Umbilicaria papulosa*); CCLS[d]
Trebouxia sp. (no lichens mentioned); CCAP 213/1b[i]
Trebouxia sp. (no lichens mentioned); Hindák 1963/33, Hindák 1963/52, Hindák 1966/9, Hindák 1967/8, Hindák 1968/39[j]
Trentepohlia sp. (from *Pyrenula nitida*); UTEX 1227
Trentepohlia sp. (from *Racodium repestre*); SAG B117.80

[a] Culture collection abbreviations: ATCC, American Type Culture Collection[58]; CCAP, Culture Centre of Algae and Protozoa[59] (updated to 1983); CCLS, Culture Collection of Lichen Symbionts at Clark University;[60] Hindák, Culture Collection of Algae at Trebon;[61] IBSG, Culture Collection of Algae at Innsbruck[62] (partially updated to 1986); PU, Culture Collection of Algae at Philipps Universitat;[63] SAG, Culture Collection of Algae at Göttingen;[64,65] UTEX, Culture Collection of Alage at University of Texas[44] (updated to 1985).
[b] Photobiont (name of lichen from which it was isolated); culture collection abbreviation and strain number.
[c] Two strains listed in Büdel[63] (1982/17136aPH, 1981/27680bPH) are not presently available.[67]
[d] Available from CCLS according to Ahmadjian.[60]
[e] *Prototheca* is not a recognized genus of phycobiont (see Volume 1, Chapter II.B).
[f] Listed in Reference 62 as *Trebouxia*.
[g] Available from CCLS according to Hildreth and Ahmadjian.[47]
[h] Not listed in the 1985 catalog.
[i] According to Reference 66, not *Trebouxia*.
[j] Strains have not been verified.

Table 3
MYCOBIONTS AVAILABLE FROM CULTURE COLLECTIONS[a,b]

Myocbiont	Culture collection[c]
Acarospora fuscata (Schrad.) Arn.	ATCC 18249
A. smaragdula (Wahlenb.) Th.Fr.	ATCC 18250
Anaptychia ciliaris (L.) Körb.	ATCC 18251
Anthracothecium albescens Müll.Arg.	ATCC 18252
Arthonia cinnabarina Wallr.	ATCC 18253
Bacidia incompta (Hook.) Anzi	ATCC 18254
Baeomyces roseus Pers.	ATCC 18255
Buellia punctata (Hoffm.) Mass. var. *polyspora* (Willey) Fink	ATCC 18256
B. stillinguiana J. Steiner	ATCC 18257
Caloplaca aurantiaca (Lightf.) Th.Fr.	ATCC 18258
Candelariella vitellina (Ehrh.) Müll.Agr.	ATCC 18268
Cetraria islandica (L.) Ach.	ATCC 18269
Cladina stellaris	CCLS
Cladonia bellidiflora (Ach.) Schaer.	ATCC 18270
C. boryi Tuck.	CCLS
C. cariosa (Ach.) Spreng.	CCLS
C. chlorophaea (Flörke ex Sommerf.) Spreng.	CCLS
C. cristatella Tuck.	ATCC 18271, CCLS
C. evansii	CCLS
C. floerkeana (Fr.) Flörke	CCLS
C. furcata (Huds.) Schrad.	CCLS
C. gracilis (L.) Willd.	ATCC 18272
C. macilenta Hoffm.	CCLS
C. nemoxyna (Ach.) Coem.	ATCC 18273
C. piedmontensis Merr.	ATCC 18274
C. pleurota (Flörke) Schaer.	CCLS
C. rangiferina (L.) Wigg.	ATCC 18275
C. subtenuis (Abb.) Evans	ATCC 18276
Dermatocarpon fluviatile (G.H.Web.) Th.Fr.	ATCC 18277
Graphis tenella Ach.	ATCC 18290
Huillia albocaerulescens	CCLS
Lecanora cinerea (L.) Sommerf.	ATCC 18292
L. dispersa (Pers.) Sommerf.	ATCC 18293
L. rubina (Vill.) Ach.	ATCC 18294
L. crustulata (Ach.) Spreng.	ATCC 18297
L. erratica Korb.	ATCC 18299
Lecidea sp.	ATCC 18296
Lepraria sp.	CCLS
Microthelia albidella Müll.Arg.	ATCC 18340
Opegrapha lichenoides Pers.	ATCC 18341
Parmelia centrifuga (L.) Ach.	ATCC 18342
P. conspersa (Ach.) Ach.	ATCC 18343
Physcia millegrana Degel.	ATCC 18346
P. stellaris (L.) Nyl.	ATCC 18347
Porina sandwichensis Magnus	ATCC 18348
Pyconthelia papillaria (Ehrh.) Duf.	CCLS
Pyrenula nitida (Weig.) Ach.	ATCC 18369
Ramalina americana Magnus	ATCC 18370
Sarcogyne simplex (Dav.) Nyl.	ATCC 18372
Stereocaulon vulcani Ach.	ATCC 18374
Umbilicaria papulosa Nyl.	ATCC 18375
Usnea florida (L.) Wigg.	ATCC 18376
Xanthoria parietina (L.) Th.Fr.	ATCC 18377

Table 3 (continued)
MYCOBIONTS AVAILABLE FROM CULTURE COLLECTIONS[a,b]

[a] Culture collection abbreviations: ATCC, American Type Culture Collection[58] CCLS, Culture Collection of Lichen Symbionts at Clark University.[60]

[b] Cultures from the Culture Collection of Mycobionts at George Mason University (quoted in Reference 60) are not presently available[68] Some of these may be obtainable from Ahmadjian.[60]

[c] Culture collection abbreviation followed by strain number.

REFERENCES

1. **Lechowicz, M. J.,** Age dependence of photosynthesis in the caribou lichen *Cladina stellaris, Plant Physiol.*, 71, 893, 1983.
2. **Larson, D. W.,** Patterns of lichen photosynthesis and respiration following prolonged frozen storage, *Can. J. Bot.*, 56, 2119, 1978.
3. **Kappen, L.,** Response to extreme environments, in *The Lichens*, Ahmadjian, V. and Hale, M. E., Eds., Academic Press, New York, 1973, 311.
4. **Kershaw, K. A.,** *Physiological Ecology of Lichens*, Cambridge University Press, London, 1985.
5. **Cowan, D. A., Green, T. G. A., and Wilson, A. T.,** Lichen metabolism. I. The use of tritium-labelled water in studies of anhydrobiotic metabolism in *Ramalina celastri* and *Peltigera polydactyla, New Phytol.*, 82, 489, 1979.
6. **Green, T. G. A. and Snelgar, W. P.,** Carbon dioxide exchange in lichens. Relationship between net photosynthetic rate and CO_2 concentration, *Plant Physiol.*, 68, 199, 1981.
7. **Chambers, S., Morris, M., and Smith, D. C.,** Lichen physiology. XV. The effect of digitonin and other treatments on biotrophic transport of glucose from alga to fungus in *Peltigera polydactyla, New Phytol.*, 76, 485, 1976.
8. **Huss-Danell, K.,** The cephalodia and their nitrogenase activity in the lichen *Stereocaulon paschale, Z. Pflanzenphysiol.*, 95, 431, 1979.
9. **Pearson, L. C.,** Varying environmental factors in order to grow intact lichens under laboratory conditions, *Am. J. Bot.*, 57, 659, 1970.
10. **Pearson, L. C. and Benson, S.,** Laboratory growth experiments with lichens based on distribution in nature, *Bryologist*, 80, 317, 1977.
11. **Fahselt, D.,** Lichen products of *Cladonia stellaris* and *Cladonia rangiferina* maintained under artificial conditions, *Lichenologist*, 13, 87, 1981.
12. **Hamada, N.,** The content of lichen substances in *Ramalina siliquosa* cultured at various temperatures in growth cabinets, *Lichenologist*, 16, 98, 1984.
13. **Kershaw, K. A. and Millbank, J. W.,** A controlled environment lichen growth chamber, *Lichenologist*, 4, 83, 1969.
14. **Kershaw, K. A. and Millbank, J. W.,** Isidia as vegetative propagules in *Peltigera aphthosa* var. *variolosa* (Massal.) Thoms., Lichenologist, 4, 214, 1970.
15. **Millbank, J. W.,** The assessment of nitrogen fixation and throughput by lichens. I. The use of a controlled environmental growth chamber to relate acetylene reduction estimates to nitrogen fixation, *New Phytol.*, 89, 647, 1981.
16. **Dibben, M. J.,** Whole-lichen culture in a phytotron, *Lichenologist*, 5, 1, 1971.
17. **Ahmadjian, V.,** Methods of isolating and culturing lichen symbionts and thalli, in *The Lichens*, Ahmadjian, V. and Hale, M. E., Eds., Academic Press, New York, 1973, 653.
18. **Galun, M., Marton, K., and Behr, L.,** A method for the cultivation of lichen thalli under controlled conditions, *Arch. Mikrobiol.*, 83, 189, 1972.
19. **Ahmadjian, V., Russell, L. A., and Hildreth, K. C.,** Artificial reestablishment of lichens. I. Morphological interactions between the phycobionts of different lichens and the mycobionts *Cladonia cristatella* and *Lecanora chrysoleuca, Myocologia*, 72, 73, 1980.
20. **Ahmadjian, V. and Jacobs, J. B.,** Relationship between fungus and alga in the lichen *Cladonia cristatella* Tuck., *Nature (London)*, 289, 169, 1981.
21. **Culberson, C. F., Culberson, W. L., and Johnson, A.,** Genetic and environmental effects on growth and production of secondary compounds in *Cladonia cristatella, Biochem. Syst. Ecol.*, 11, 77, 1983.

22. **Drew, E. A. and Smith, D. C.,** The physiology of the symbiosis in *Peltigera polydactyla* (Neck.) Hoffm., *Lichenologist,* 3, 197, 1966.
23. **Drew, E. A. and Smith, D. C.,** Studies in the physiology of lichens. VII. The physiology of the *Nostoc* symbiont of *Peltigera polydactyla* compared with cultured and free-living forms, *New Phytol.,* 66, 379, 1967.
24. **Richardson, D. H. S.,** Lichens, in *Methods in Microbiology,* Vol. 4, Booth, C., Ed., Academic Press, New York, 1971, 267.
25. **Richardson, D. H. S. and Smith, D. C.,** Lichen physiology. X. The isolated algal and fungal symbionts of *Xanthoria aureola, New Phytol.,* 67, 69, 1968.
26. **Ahmadjian, V.,** *The Lichen Symbiosis,* Blaisdell, Waltham, 1967.
27. **Bubrick, P. and Galun, M.,** Symbiosis in lichens: differences in cell wall properties of freshly isolated and cultured phycobionts, *FEMS Microbiol. Lett.,* 7, 311, 1980.
28. **Ascaso, C.,** A rapid method for the quantitative isolation of green algae from lichens, *Ann. Bot.,* 45, 483, 1980.
29. **Galun, M. and Bubrick, P.,** Physiological interactions between the partners of the lichen symbiosis, in *Encyclopedia of Plant Physiology,* Vol. 17, Linskens, H. F. and Heslop-Harrison, J., Eds., Springer-Verlag, Berlin, 1984, 362.
30. **Honegger, R. and Brunner, U.,** Sporopollenin in the cell walls of *Coccomyxa* and *Myrmecia* phycobionts of various lichens: an ultrastructural and chemical investigation, *Can. J. Bot.,* 59, 2713, 1981.
31. **Perez, M. J., Vicente, C., and Legaz, M. E.,** An improved method to isolate lichen algae by gel filtration, *Plant Cell Rep.,* 4, 210, 1985.
32. **Ahmadjian, V.,** A guide to the algae occurring as lichen symbionts: isolation, culture, physiology, and identification, *Phycologia,* 6, 127, 1967.
33. **Stein, J. R., Ed.,** *Handbook of Phycological Methods. Culture Methods and Growth Measurements,* Cambridge University Press, London, 1973.
34. **Bubrick, P. and Galun, M.,** Cyanobiont diversity in the Lichinaceae and Heppiaceae, *Lichenologist,* 16, 279, 1984.
35. **Büdel, B. and Henssen, A.,** *Chroococcidiopsis* (Cyanophyceae), a phycobiont in the lichen family Lichinaceae, *Phycologia,* 22, 367, 1983.
36. **Kratz, W. A. and Myers, J.,** Nutrition and growth of several blue-green algae, *Am. J. Bot.,* 42, 282, 1955.
37. **Gorham, P. R., McLachlan, J. S., Hammer, V. T., and Kim, W. K.,** Isolation and culture of toxic strains of *Anabaena flos-aquae* (Lyngb.) de Breb., *Verh. Int. Verein. Theor. Angew. Limmol.,* 15, 796, 1964.
38. **Stanier, R. Y., Kunisawa, R., Mandel, M., and Cohen-Bazire, G.,** Purification and properties of unicellular blue-green algae (order Chroococcales), *Bact. Rev.,* 35, 171, 1971.
39. **Allen, M. M.,** Simple conditions for growth of unicellular blue-green algae, *J. Phycol.,* 4, 1, 1968.
40. **Allen, M. M.,** Methods for Cyanophyceae, in *Handbook of Phycological Methods. Culture Methods and Growth Measurements,* Stein, J. R., Ed., Cambridge University Press, London, 1973, 127.
41. **Carr, N. G., Komarek, J., and Whitton, B. A.,** Notes on isolation and laboratory culture, in *The Biology of Blue-Green Algae,* Carr, N. G. and Whitton, B. A., Eds., University of California Press, Berkeley, 1973, 523.
42. **Nichols, H. W. and Bold, H. C.,** *Trichosarcina polymorpha* gen. et sp. nov., *J. Phycol.,* 1, 34, 1965.
43. **Nichols, H. W** Growth media—fresh water, in *Handbook of Phycological Methods. Culture Methods and Growth Measurements,* Stein, J. R., Ed., Cambridge University Press, London, 1973, 7.
44. **Starr, R. C.,** The culture collection of algae at the University of Texas at Austin, *J. Phycol.,* 14 (Suppl.), 483, 1980.
45. **Archibald, P. A.,** Physiological characteristics of *Trebouxia* (Chlorophyceae, Chlorococcales) and *Pseudotrebouxia* (Chlorophyceae, Chlorosarcinales), *Phycolgia,* 16, 295, 1977.
46. **Archibald, P. A.,** *Trebouxia* de Puymaly (Chlorophyceae, Chlorococcales) and *Pseudotrebouxia* gen. nov. (Chlorophyceae, Chlorosarcinales), *Phycologia,* 14, 125, 1975.
47. **Hildreth, K. C. and Ahmadjian, V.,** A study of *Trebouxia* and *Pseudotrebouxia* isolates from different lichens, *Lichenologist,* 13, 65, 1981.
48. **Ahmadjian, V.,** Qualitative requirements and utilization of nutrients: lichens, in *CRC Handbook Series of Nutrition and Food, Section D, Nutritional Requirements,* Vol. 1. Rechcigl, M., Jr., Ed., CRC Press, Boca Raton, Fla., 1977, 203.
49. **Lockart, C. M., Rowell, P., and Stewart, W. D. P.,** Phytohaemagglutinins from the nitrogen-fixing lichens *Peltigera canina* and *P. polydactyla, FEMS Microbiol. Lett.,* 3, 127, 1978.
50. **Pyatt, F. B.,** Lichen propagules, in *The Lichens,* Ahmadjian, V. and Hale, M. E., Eds., Academic Press, New York, 1973, 117.
51. **Scott, G. D.,** Observations on spore discharge and germination in *Peltigera praetextata* (Flk.) Vain., *Lichenologist,* 1, 109, 1959.

52. **Lallement, R. and Bernard, T.,** Obtention de cultures pures des mycosymbiotes du *Lobaria laetevirens* (Lightf.) Zahlbr. et du *Lobaria pulmonaria* (L.) Hoffm.: le role des gonidies, *Rev. Bryol. Lichenol.*, 43, 303, 1977.
53. **Galun, M., Braun, A., Frensdorff, A., and Galun, E.,** Hyphal walls of isolated lichen fungi. Autoradiographic localization of precursor incorporation and binding of fluorescein-conjugated lectins, *Arch. Microbiol.*, 108, 9, 1976.
54. **Bubrick, P., Galun, M., and Frensdorff, A.,** Proteins from the lichen *Xanthoria parietina* which bind to phycobiont cell walls. Localization in the intact lichen and cultured myocobiont, *Protoplasma*, 105, 207, 1981.
55. **Vobis, G.,** Studies on the germination of lichen conidia *Lichenologist*, 9, 131, 1977.
56. **Bianchi, E. E.,** An endogenous circadian rhythm in *Neurospora crassa, J. Gen. Microbiol.*, 35, 437, 1964.
57. **Lilly, V. G. and Barnett, H. L.,** *Physiology of the Fungi*, McGraw-Hill, New York, 1951.
58. **Jong, S. C. and Gantt, M. J.,** *American Type Culture Collection. Catalog of Fungi/Yeasts*, 17th ed., American Type Culture Collection, Rockville, Md., 1987.
59. **George, E. A., Ed.,** List of strains, *Culture Centre of Algae and Protozoa*, Cambridge, 1976.
60. **Ahmadjian, V.,** Guide to culture collections of lichen symbionts, *Int. Lichenol. Assoc. News*. 13, 13, 1980.
61. **Hindák, F.,** Culture collection of algae at Laboratory of Algology in Trebon, *Arch. Hydrobiol. Suppl.*, 2/3 (Suppl. 39), 86, 1970.
62. **Gártñer, G.,** Verzeichnis der Algenkulturen am Institut für Botanische Systematik und Geobotanik der Universitat Innsbruck, *Ber. Nat. Med. Ver. Innsbruck*, 63, 67, 1976.
63. **Büdel, B.,** Blue-green phycobionts in the family Lichinaceae, *Arch. Hydrobiol. Suppl.*, 38/39 (Suppl. 71), 355, 1985.
64. **Schlósser, U. G.,** Sammlung von Algenkulturen, *Ber. Dtsch. Bot. Ges.*, 95, 181, 1982.
65. **Schlösser, U. G.,** Sammlung von Algenkulturen Göttingen: additions to the collection since 1982, *Ber. Dtsch. Bot. Ges.*, 97, 465, 1984.
66. **Bubrick, P.,** unpublished observations.
67. **Büdel, B.,** personal communication.
68. **Lawrey, J. D.,** personal communciation.

Index

INDEX

D

K

L

M

N

O

P

Q

R

S

T

U

V

W

X